BASIC heat transfer

Butterworths BASIC Series includes the following titles:

BASIC aerodynamics
BASIC artificial intelligence
BASIC business analysis and operations research
BASIC business systems simulation
BASIC differential equations
BASIC digital signal processing
BASIC economics
BASIC electrotechnology
BASIC forecasting techniques
BASIC fluid mechanics
BASIC graph and network algorithms
BASIC heat transfer
BASIC hydraulics
BASIC hydrodynamics
BASIC hydrology
BASIC interactive graphics
BASIC investment appraisal
BASIC materials studies
BASIC matrix methods
BASIC mechanical vibrations
BASIC molecular spectroscopy
BASIC numerical mathematics
BASIC operational amplifiers
BASIC reliability engineering analysis
BASIC soil mechanics
BASIC statistics
BASIC stress analysis
BASIC surveying
BASIC technical systems simulation
BASIC theory of structures
BASIC thermodynamics and heat transfer
BASIC water and waste water treatment

BASIC heat transfer

D. H. Bacon MSc, BSc(Eng), CEng, MIMechE
Polytechnic South West

Butterworths
London Boston Singapore Sydney Toronto Wellington

First published in 1989

British Library Cataloguing in Publication Data
Bacon, D. H. (Dennis Henry), *1930–*
BASIC heat transfer.
1. Heat transfer. Analysis. Applications of computer systems
I. Title
536′.2′00285

ISBN 0-408-01275-7

Library of Congress Cataloging-in-Publication Data
Bacon, D. H. (Dennis Henry)
Basic heat transfer / D. H. Bacon
p. cm.
Includes bibliographical references.
ISBN 0-408-01275-7
1. Heat-Transmission-Data processing. 2. BASIC (Computer program language) I. Title.
TJ260.B28 1990
621.402′2′02855133-dc20

Photoset by The Alden Press London, Northampton and Oxford
Printed and bound by Hartnolls Ltd., Bodmin, Cornwall

Preface

BASIC heat transfer is not intended as a comprehensive treatise of BASIC or heat transfer. It aims to help readers to use a computer to solve heat transfer problems and to promote greater understanding by changing data values and observing the effects. Such steps are necessary in design and optimisation calculations.

Traditional heat transfer books are usually lengthy and contain a considerable amount of fundamental analysis. This book does not reproduce this standard work but gives references which may be used if the information is required. Essentially the book is concerned with applications including insulation and heating in buildings and pipes, temperature distributions in solids for steady state and transient conditions, the determination of surface heat transfer coefficients for convection in various situations, radiation heat transfer in grey body problems, the use of finned surfaces, and simple heat exchanger design calculations. The theory and worked examples cover these fields. The problems revise and extend the work and, additionally, each chapter has at least one open-ended problem which requires further background reading and investigation by the reader to extend his knowledge of heat transfer.

Chapter 1 includes a review of the BASIC computing required and also includes some mathematical programs which will be required later to solve heat transfer problems. The book also contains a number of graphs from which data can be taken. These have been plotted from equations included in the text which may be programmed by the reader if required. The graphs are only of sufficient accuracy for use in this book. All the symbols used are defined as they first appear but definitions of those in common use are not repeated.

The programs presented are straightforward without confusing complications and the liberal use of REM statements coupled with the program notes which include an explanation of the structure should enable the reader to follow a program without a flow chart. However, charts could be drawn if required to clarify any points and, indeed, could be used to streamline some of the steps. If the reader chooses, the use of such charts might help in the solution of some of the problems presented.

D.H.B.

Acknowledgement

The author thanks Séan Cole for his last-minute assistance.

Contents

Chapter 1

Introduction to BASIC

1.1 The BASIC approach

The programs in this book are written in the BASIC language. This chapter is not an instruction manual in BASIC but, rather, a short description of the simple BASIC used in this book.

1.2 The elements of BASIC

1.2.1 Mathematical expressions

One of the main objects of the example programs in this book is to assist in the learning of BASIC by applying it to a relevant engineering subject. This aim can be met by the reader studying the examples, using them and then trying some of the problems. It will be necessary to evaluate equations which contain numerical constants, variables (e.g. x) and functions (i.e. sine). All numbers are treated identically whether they are integer (e.g. 36) or real (e.g. 36.1). An exponential form is used to represent large or small numbers (e.g. 3.61E6 which equals 3.61×10^6). Numeric variables are represented by combinations of letters or letters and numbers (e.g. T, TT, T1, T2E). For generality π is always written as 3.142 in this book. The determination of square roots uses an in-built function (e.g. SQR(X)) but in this book the alternative (X ∧ 0.5) is used. The argument in brackets (X) can be a number, a variable or a mathematical expression. For trigonometric functions (SIN(X), COS(X), etc.) the argument is interpreted as being measured in radians. Other functions include a natural logarithm (LOG) and its exponential (EXP) and ABS which gives the absolute value of the argument.

Mathematical equations also contain operators such as plus and minus, etc. These operators have a hierarchy in that some are performed by the computer before others. In descending order of priority the operators are:

(a) to the power of (∧);
(b) multiply (*) and divide (/);
(c) add (+) and subtract (−).

Thus, for example, multiplication is done before addition. The computer works from left to right if the operators have the same priority. Brackets can be used to override any of these operations. Hence

$$\frac{a + b}{3c} \quad \text{becomes} \quad (A + B)/(3 * C) \quad \text{or} \quad (A + B)/3/C$$

1.2.2 Program structure and assignment

A BASIC program is a sequence of statements which define a procedure for the computer to follow. As it follows this procedure the computer allocates values to each of the variables. The values of some of these variables may be specified by data that is input to the program. Others are generated in the program using, for instance, the assignment statement. This has the form

line number [LET] variable = mathematical expression

where the word LET is usually optional and therefore omitted. For example, the root of a quadratic equation

$$x_1 = \frac{-b + (b^2 - 4ac)^{1/2}}{2a}$$

may be obtained from a statement such as

100 X1 = (– B + SQR(B ∧ 2 – 4 * A * C))/(2 * A)

It is important to realise that an assignment statement is not itself an equation. It is an instruction to give the variable on the left-hand side the numeric value of the expression on the right-hand side. Thus it is possible to have a statement

50 X = X + 1

which increases by 1 the value of X. Each variable can have only one value at any time unless it is subscripted (see Section 1.2.7).

Note that all BASIC statements (i.e. all the program lines) are numbered. This defines the order in which they are executed.

1.2.3 Input

For 'interactive' or 'conversational' programs the user specifies values for variables in response to questions displayed on the screen. The statement has the form

line number INPUT variable 1 [, variable 2, . . .]

e.g.

```
20 INPUT A, B, C
```

When the program is run the computer prints ? as it reaches this statement and waits for the user to type values for the variables, e.g.

```
? 5 ? 10 ? 15
```

which makes A = 5, B = 10 and C = 15 in the above example.

An alternative form of data input is useful if there are many data or if the data are not to be changed by the user (e.g. the water properties data). For this type of data specification there is a statement of the form

line number READ variable 1 [, variable 2, . . .]

e.g.

```
20 READ A, B, C
```

with an associated statement (or number of statements) of the form

line number DATA number 1 [, number 2, . . .]

e.g.

```
1 DATA 5, 10, 15
```

or

```
1 DATA 5
2 DATA 10
3 DATA 15
```

DATA statements can be placed anywhere in the program – it is often convenient to place them at the beginning or end so that they can be easily changed.

1.2.4 Output

Output of data and the results of calculations, etc., is done using a statement of the form

line number PRINT list

The list may contain variables or expressions, e.g.

```
200 PRINT A, B, C, A * B/C
```

text enclosed in quotes, e.g.

```
10 PRINT "INPUT A, B, C IN MM";
```

or mixed text and variables, e.g.

```
300 PRINT "PRESSURE="; P; "KN/M ∧ 2"
```

The items in the list are separated by commas or semi-colons. Commas give tabulation in columns, each about 15 spaces wide. A semi-colon suppresses this spacing and if it is placed at the end of a list it suppresses the line feed. If the list is left unfilled a blank line is printed; TAB statements may also be used to arrange output format.

Note the necessity to use PRINT statements in association with both 'run-time' input (to indicate what input is required) and READ/DATA statements (because otherwise the program user has no record of the data).

1.2.5 Conditional statements

It is often necessary to enable a program to take some action if, and only if, some condition is fulfilled. This is done with a statement of the form

line number IF expression 1 conditional operator expression 2 THEN GOTO line number

where the possible conditional operators are

= equals
< > not equal to
< less than
< = less than or equal to
> greater than
> = greater or equal to

For example, a program could contain the following statements if it is to stop when a zero value of A is input:

```
20 INPUT A
30 IF A < > 0 THEN GOTO 50
40 STOP
50 . . .
```

Note the statement

line number STOP

which stops the run of a program.

1.2.6 Loops

There are several means by which a program can repeat some of its actions. The simplest such statement is

line number GO TO line number

This can be used, for instance, with the conditional statement example above so that the program continues to request values of A until the user inputs zero.

The most common means for performing loops is with a starting statement of the form

line number FOR variable = expression 1 TO expression 2 [STEP expression 3]

where the STEP is assumed to be unity if omitted. The finish of the loop is signified by a statement:

line number NEXT variable

where the same variable is used in both FOR and NEXT statements. Its value should not be changed in the intervening lines.

A loop is used if, for example, N sets of data have to be READ and their reciprocals printed, e.g.

```
10 READ N
20 PRINT "NUMBER", "RECIPROCAL"
30 FOR I = 1 TO N
40 READ A
50 PRINT A, 1/A
60 NEXT I
```

1.2.7 Subscripted variables

It is sometimes very beneficial to allow a single variable to have a number of different values during a single program run. For instance if a program contains data for several materials it is convenient for their densities to be called R(1), R(2), R(3), etc., instead of R1, R2, R3, etc. It is then possible for a single statement to perform calculations for all the materials, e.g.

```
50 FOR I + 1 TO N
60 M(I) = V * R(I)
70 NEXT I
```

which determines the mass M(I) for each material from the volume (V) of the body.

A non-subscripted variable has a single value associated with it but if a subscripted variable is used it is necessary to provide space in the memory for all the values. This is done with a dimensioning statement of the form line number DIM variable 1 (integer 1) [, variable 2 (integer 2), . . .]

e.g.

20 DIM R(50), M(50)

which allows up to 50 values of R amd M. The DIM statement must occur before the subscripted variables are first used and it usually appears early in the program.

1.2.8 Subroutines

Sometimes a sequence of statements needs to be accessed more than once in the same program. Instead of merely repeating these statements it is better to put them in a subroutine. The program then contains statements of the form

line number GOSUB line number

When the program reaches this statement it branches (i.e. transfers control to) to the second line number. The sequence of statements comprising the subroutine starting with this second line number ends with a statement:

line number RETURN

and the program returns control to the statement immediately after the GOSUB call.

Subroutines can be placed anywhere in the program but it is usually convenient to position them at the end, separate from the main program statements.

1.2.9 Other statements

1. Explanatory remarks or headings which are not to be output can be inserted into a program using

   line number REM comment

   Any statement beginning with the word REM is ignored by the computer.
2. The integer part of a number may be obtained with the INT function. This function can also be used to round up a number or to reduce the number of decimal places displayed or printed by the method given below.

   line number variable = (INT(variable * 100 + 0.5))/100

   The multiplier and divisor 100 may be 10 to any power.

1.3 Checking programs

Most computers give a clear indication if there are grammatical (syntax) errors in a BASIC program. Program statements can be modified by retyping them correctly or by using special editing procedures. The majority of syntax errors are easy to locate but if a variable has been used with two (or more) different meanings in separate parts of the program, some mystifying errors can result.

It is not sufficient for the program to be just grammatically correct. It must also give the correct answers. A program should therefore be checked either by using data which gives a known solution or by hand calculation. If the program is to be used with a wide range of data or by users other than the program writer it is necessary to check that all parts of it function. It is also important to ensure that the program does not give incorrect but plausible answers when 'nonsense' data is input. It is quite difficult to make programs completely 'userproof' and they often become lengthy in doing so. The programs in this book have been kept as short as possible for the purpose of clarity and may not therefore be fully 'userproof'.

1.4 Summary of BASIC statements

Assignment	
DIM	Allocates space for subscripted variables
Input	
INPUT	Interactive input of data from keyboard
READ	Reads data from DATA statements
DATA	Storage area for data
Output	
PRINT	Prints output
TAB	Arranges spacing of printed output
Program control	
STOP or END	Stops program run
GOTO	Unconditional branching
IF . . . THEN GOTO	Conditional branching
FOR . . . TO . . . STEP	Opens loop
NEXT	Closes loop
GOSUB	Transfers control to subroutine
RETURN	Returns control from subroutine
Comment	
REM	Comment in program

Functions

SQR	Square root
SIN	Sine (angle in radians)
COS	Cosine (angle in radians)
LOG	Natural logarithm (base e)
EXP	Exponential
ABS	Absolute value
INT	Integer part

References

1. Alcock, D., *Illustrating BASIC*, Cambridge University Press (1988).
2. Monro, D.M., *Interactive Computing with BASIC*, Edward Arnold (1977).

WORKED EXAMPLES

Program 1.1 Useful mathematical routines

Three useful routines which will be required in later programs follow:

1. *Lagrangian interpolation* is based on the formation of an nth order polynomial from $(n + 1)$ points, which need not be evenly spaced. The data for interpolation has to be read into the program and a trivial demonstration is included in the program RUN.
2. *Gaussian elimination* enables the solution of a number of simultaneous linear equations. The number of coefficients is equal to the number of equations and there is a constant for each:

$$a_1x_1 + a_2x_2 + a_3x_3 \ldots a_nx_n = C_1$$

$$b_1x_1 + b_2x_2 + b_3x_3 \ldots b_nx_n = C_2$$

Thus the program for n equations needs n sets of $(n + 1)$ pieces of data. Trivial examples are included in the program RUN.

3. *Straight line fit* enables the equation to a straight line to be determined by the method of least squares. The data required is n sets of two coordinates. A trivial example is included in the program RUN.

```
10  PRINT "LAGRANGIAN INTERPOLATION TO FIND H FOR"
20  PRINT "A GIVEN T FROM A SET OF (T,H) DATA"
30  DIM T(30),H(30)
40  FOR N = 0 TO 4
50  READ T(N),H(N)
60  NEXT N
```

```
70  PRINT " WHAT IS THE VALUE OF T AT WHICH H IS REQUIRED
?"
80  INPUT A: PRINT " FIND H AT T= ";A
1000 B = 0
1010  FOR J = 0 TO 4
1020 X = 1
1030  FOR N = 0 TO 4
1040  IF N = J THEN  GOTO 1060
1050 X = X * (A - T(N)) / (T(J) - T(N))
1060  NEXT N
1070 B = B + X * H(J)
1080  NEXT J
1090  PRINT " AT T=";A;" THE VALUE OF H=";B
1100  REM  THE DATA FOR INTERPOLATION MUST BE AVAILABLE F
OR THE PROGRAM TO OPERATE
1200  DATA  1,2
1210  DATA  2,4
1220  DATA  3,6
1230  DATA  4,8
1240  DATA  6,12

]RUN
LAGRANGIAN INTERPOLATION TO FIND H FOR
A GIVEN T FROM A SET OF (T,H) DATA
 WHAT IS THE VALUE OF T AT WHICH H IS REQUIRED?
?5
 FIND H AT T= 5
 AT T=5 THE VALUE OF H=10
```

```
10  PRINT "GAUSSIAN ELIMINATION PROGRAM"
15  REM   BEFORE RUNNING THIS PROGRAM IT IS ESSENTIAL TO
ADD THE DATA STATEMENTS WITH THE COEFFICIENTS AND CONSTAN
TS OF THE EQUATIONS TO BE SOLVED(ONE LINE FOR EACH EQUATI
ON)
20  PRINT
30  DIM A(30,31): DIM X(30)
40  PRINT "NUMBER OF EQUATIONS";
50  INPUT R: PRINT " SOLVING ";R;" EQUATIONS"
60  FOR J = 1 TO R
70  FOR I = 1 TO R + 1
80  READ A(J,I)
90  NEXT I
100  NEXT J
170  FOR J = 1 TO R
180  FOR I = J TO R
190  IF A(I,J) <  > 0 THEN  GOTO 230
200  NEXT I
210  PRINT " SOLUTION NOT POSSIBLE"
220  GOTO 1000
230  FOR K = 1 TO R + 1
240 X = A(J,K)
250 A(J,K) = A(I,K)
260 A(I,K) = X
270  NEXT K
280 Y = 1 / A(J,J)
290  FOR K = 1 TO R + 1
300 A(J,K) = Y * A(J,K)
310  NEXT K
320  FOR I = 1 TO R
330  IF I = J THEN  GOTO 380
340 Y =  - A(I,J)
350  FOR K = 1 TO R + 1
360 A(I,K) = A(I,K) + Y * A(J,K)
370  NEXT K
380  NEXT I
390  NEXT J
```

```
400  PRINT
405  PRINT "SOLUTION IS"
410  FOR I = 1 TO R
420 X(I) =  INT (A(I,R + 1) * 100 + 0.5) / 100
430  NEXT I
440  FOR I = 1 TO R
450  PRINT "X";I;"= ";X(I)
460  NEXT I
1000  END

500 DATA 1,2,4,7

510 DATA 2,4,8,14

520 DATA 4,8,16,28

]RUN
GAUSSIAN ELIMINATION PROGRAM

NUMBER OF EQUATIONS?3
 SOLVING 3 EQUATIONS
 SOLUTION NOT POSSIBLE

500 DATA 3,2,1,11

510 DATA 2,4,8,14

520 DATA 1,2,-1,1

]RUN
GAUSSIAN ELIMINATION PROGRAM

NUMBER OF EQUATIONS?3
 SOLVING 3 EQUATIONS

SOLUTION IS
X1= 3.8
X2= -.8
X3= 1.2
```

```
3000  PRINT "STRAIGHT LINE FIT"
3010  DIM X(20),Y(20)
3020  REM  THE PROGRAM REQUIRES DATA FOR N VALUES OF X AN
D Y(N<=20)
3030  PRINT "WHAT IS THE NUMBER OF POINTS ON THE LINE?"
3040  INPUT N: PRINT "THERE ARE ";N;" POINTS"
3050  FOR Z = 1 TO N
3060  READ X(Z),Y(Z)
3070  NEXT Z
3100 A1 = 0:B1 = 0:C1 = 0:D1 = 0
3110  FOR Z = 1 TO N
3120 A1 = A1 + X(Z)
3130 B1 = B1 + Y(Z)
3140 C1 = C1 + (X(Z) ^ 2)
3150 D1 = D1 + (X(Z) * Y(Z))
3160  NEXT Z
3170 F1 = ((N * D1) - (A1 * B1)) / ((N * C1) - (A1 ^ 2))
3180 G1 = (B1 - (F1 * A1)) / N
3190  PRINT "REQUIRED EQUATION IS Y=";F1;"*X+";G1
3200  REM  ROUNDING OFF TERMS MAY BE ADDED IF REQUIRED
4000  DATA  1,5
4010  DATA  2,6.9
4020  DATA  3,9.1
4030  DATA  4,10.9
4040  DATA  5,13
4050  DATA  6,15.1
```

```
]RUN
STRAIGHT LINE FIT
WHAT IS THE NUMBER OF POINTS ON THE LINE?
?6
THERE ARE 6 POINTS
REQUIRED EQUATION IS Y=2.01714287*X+2.93999996
```

Chapter 2

Simple heat transfer

2.1 Introduction

Whenever two systems are not in thermal equilibrium due to a temperature difference between them, energy is transferred from the hot system to the cold system in an attempt to achieve equilibrium. There are three modes of heat transfer: conduction, convection and radiation. In real situations all three modes occur simultaneously and a balance is drawn up to determine the total effect. For example, in an electric storage heater the energy supplied to the element is transferred by radiation (with small contributions from conduction and convection) to the storage bricks. The energy is then conducted to the metal surface of the heater from which it is convected and radiated to the room. The total effect is, of course, that the energy supplied to the element is all transferred to the room, albeit with a time delay. Before such problems can be considered the fundamental mechanism and laws of conduction, convection and radiation must be established.

2.2 Conduction

Conduction is an energy transfer due to interaction between molecules at different temperatures. It occurs in solids, liquids and gases but is of more consequence in solids which have high molecular densities. The fundamental law of conduction is due to Fourier:

$$\dot{Q}'' = -k \frac{\partial T}{\partial n}$$

where $\dot{Q}''$ is the heat transfer rate per unit area in the n direction, k is the thermal conductivity which is a function of temperature and $\partial T/\partial n$ is the temperature gradient in the n direction.

2.3 Thermal conductivity

Although k is a function of temperature it is convenient for simple calculations *in solid materials* to consider k a constant over small temperature ranges. This technique is not suited to liquids and gases and the functional relationship or tabulations of data will be required for solution of problems. The unit of thermal conductivity is W/m K.

2.4 One-dimensional conduction

For one-dimensional situations, Fourier's law becomes

$$\dot{Q}'' = -k\frac{dT}{dx} \tag{2.1}$$

This can be integrated for plane and tubular walls.

For a plane wall (Figure 2.1(a))

$$\dot{Q}''dx = -k\,dT$$

Integrate

$$\dot{Q}''(x_2 - x_1) = -k(T_2 - T_1)$$

$$\dot{Q}'' = \frac{-k(T_2 - T_1)}{(x_2 - x_1)} = \frac{\Delta T}{\left(\dfrac{\Delta x}{k}\right)} \tag{2.2}$$

where ΔT is the temperature difference across the wall and Δx is the wall thickness.

For a tubular wall (Figure 2.1(b)) it is more convenient to write Fourier's law in terms of heat transfer rate per unit *length* ($\dot{Q}'$):

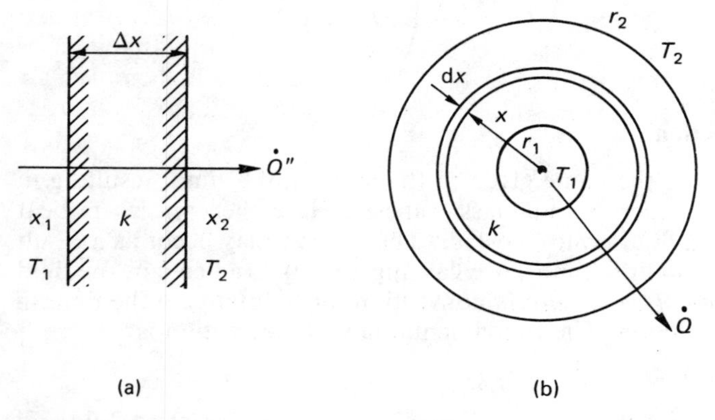

Figure 2.1

$$\dot{Q}' = -k2\pi\, x \frac{\mathrm{d}T}{\mathrm{d}x} \tag{2.3}$$

$$\frac{\dot{Q}'}{2\pi k}\left(\frac{\mathrm{d}x}{x}\right) = -\mathrm{d}T$$

Integrate

$$\frac{\dot{Q}'}{2\pi k}\ln\left(\frac{r_2}{r_1}\right) = -(T_2 - T_1)$$

$$\dot{Q}' = \frac{-2\pi k(T_2 - T_1)}{\ln\left(\frac{r_2}{r_1}\right)} = \frac{\Delta T}{\left(\frac{\ln\left(\frac{r_2}{r_1}\right)}{2\pi k}\right)} \tag{2.4}$$

Similar integration can be applied to each layer of a multilayer wall or tube (an example is the application of insulation) to give (where ΔT is the *overall* temperature difference across the wall):

For a plane multilayer wall

$$\dot{Q}'' = \frac{\Delta T}{\sum\left(\frac{\Delta x}{k}\right)} \tag{2.5}$$

For a tubular multilayer wall

$$\dot{Q}' = \frac{\Delta T}{\sum\left(\frac{\ln\left(\frac{r_2}{r_1}\right)}{2\pi k}\right)} \tag{2.6}$$

2.5 Convection

Convection is the name given to the motion of a fluid, resulting in momentum, energy and mass transfer. Here the concern is heat transfer at a fluid–solid interface. The motion may occur as a result of conduction at the interface causing density gradients in the fluid which causes *free or natural* convection, or by pumping the fluid in *forced* convection. The fundamental law of convection is

$$\dot{Q}'' = h\theta$$

where θ is the temperature difference between surface and fluid and

h is the surface heat transfer coefficient in $W/m^2 K$. The determination of h, which is not a constant, is complex (Chapter 4) and depends on fluid properties, the flow pattern and the boundary shape.

For a plane surface

$$\dot{Q}'' = h\theta = \frac{\theta}{\left(\frac{1}{h}\right)} \tag{2.7}$$

For a tubular surface

$$\dot{Q}' = 2\pi r h\theta = \frac{\theta}{\left(\frac{1}{2\pi r h}\right)} \tag{2.8}$$

2.6 Electrical analogy

An analogy for conduction and convection may be made with electrical conduction which is governed by Ohm's law. We consider ΔT analogous to V (the potential) and $\dot{Q}''$ or $\dot{Q}'$ analogous to I (the flow). It can then be seen from equations (2.2), (2.4), (2.7) and (2.8) that for plane surfaces $\Delta x/k$ is the thermal resistance to conduction and for tubular surfaces

$$\left(\frac{\ln\left(\frac{r_2}{r_1}\right)}{2\pi k}\right)$$

is the thermal resistance to conduction. Similarly $1/h$ and $1/2\pi rh$ are the thermal resistances to convection. Thermal resistances can be added in the same way as electrical resistances so that for a multilayer problem equations (2.5) and (2.6) can be written directly without integration.

2.7 Overall heat transfer coefficient

In problems where two fluids are separated by a solid wall the resistances for conduction and convection in series may be added and the heat transfer rate conveniently written in terms of the *fluid* temperatures on either side of the wall. In a building these would be the outside and inside air temperatures.

Referring to Figure 2.2 and putting $(T_{f_1} - T_{f_2}) = \theta$, the overall temperature difference, it can be seen that for a *plane wall*:

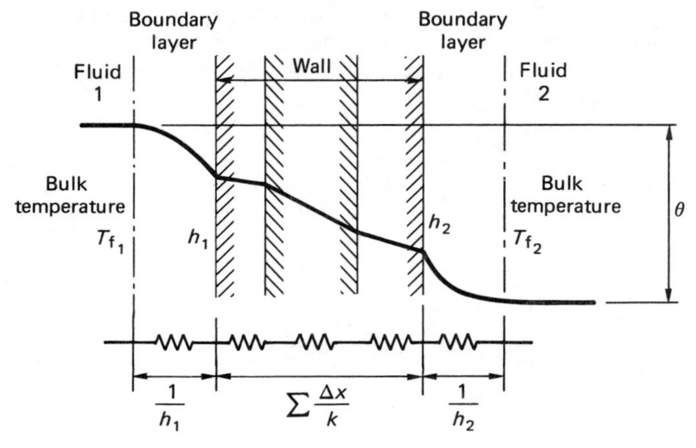

Figure 2.2

$$\dot{Q}'' = \frac{\theta}{\frac{1}{h_1} + \sum\left(\frac{\Delta x}{k}\right) + \frac{1}{h_2}}$$

An overall heat transfer coefficient U is defined by

$$\frac{1}{U} = \frac{1}{h_1} + \sum\left(\frac{\Delta x}{k}\right) + \frac{1}{h_2} \tag{2.9}$$

so that

$$\dot{Q}'' = U\theta \quad \text{or} \quad \dot{Q} = UA\theta \tag{2.10}$$

where A is the wall area and $\dot{Q}$ is the heat transfer rate.

Similarly for the *tubular* wall for which an overall heat transfer coefficient U' is defined by

$$\frac{1}{U'} = \frac{1}{2\pi r_1 h_1} + \sum\left(\frac{\ln\left(\frac{r_2}{r_1}\right)}{2\pi k}\right) + \frac{1}{2\pi r_2 h_2} \tag{2.11}$$

it is seen that

$$\dot{Q}' = U'\theta \quad \text{or} \quad \dot{Q} = U'L\theta \tag{2.12}$$

where L is the tube length and $\dot{Q}$ is the heat transfer rate. Values for U and U' are available for common constructions such as house walls, double glazed windows, etc [1].

2.8 Radiation

Radiation is the name given to the mode of heat transfer due to electromagnetic waves. No contact between transmitter and receiver is required and radiation occurs in a vacuum as well as when a gaseous media fills the space between transmitter and receiver. Any surface that is within 'sight' of any other surface will both emit radiation to it and receive radiation from it and the heat transfer rate will be the algebraic sum of these quantities. Incident radiation may be reflected, transmitted or absorbed. If ρ = reflectivity, τ = transmissivity and α = absorbtivity, then

$$\rho + \tau + \alpha = 1 \tag{2.13}$$

For solids and liquids it is usual to assume $\tau = 0$ and for gases $\rho = 0$ and $\tau = 1$ is usually assumed.

2.9 Laws of black body radiation

For ideal analysis a perfect emitter and receiver is considered.

A *black body* is defined as a surface with absorbtivity of unity.

The radiation flux per unit area $\dot{E}''$ of any body is called the *emissive power*. For a black body the Stefan–Boltzmann law states

$$\dot{E}''_{\mathrm{b}} = \sigma T^4 \tag{2.14}$$

where σ is the Stefan–Boltzmann constant, $5.67 \times 10^{-8}\,\mathrm{W/m^2\,K^4}$, and T is the thermodynamic temperature of the body. This radiation is spread over a wide spectrum of wavelengths (Figure 2.3) given by Planck's law:

$$\dot{E}''_{\mathrm{b}_\lambda} = \frac{C_1 \lambda^{-5}}{\exp\left(\dfrac{C_2}{\lambda t}\right) - 1} \tag{2.15}$$

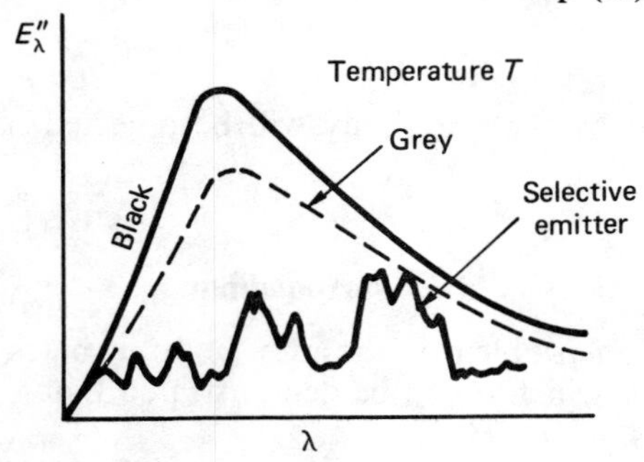

Figure 2.3

where $\dot{E}''_{b_\lambda}$ is the emissive power at wavelength λ and the constants $C_1 = 3.7415 \times 10^{-16}\,\mathrm{W\,m^2}$ and $C_2 = 1.4388 \times 10^{-2}\,\mathrm{m\,K}$.

The maximum emissive power occurs at a particular wavelength which varies with temperature according to Wien's displacement law:

$$\lambda_{max} T = 2.9 \times 10^{-3}\,\mathrm{m\,K} \quad (2.16)$$

2.10 Emissivity and grey body radiation

Real surfaces are not black and emit less radiation than black surfaces. The monochromatic emissivity ε_λ of a surface is the ratio of the emissive power of the surface at wavelength λ to the emissive power of a black body at the same wavelength, both being measured at the same temperature T:

$$\varepsilon_\lambda = \left[\frac{\dot{E}''_\lambda}{\dot{E}''_{b_\lambda}}\right]_T$$

Monochromatic emissivity of a real surface is not constant and varies considerably with λ; such surfaces are called *selective emitters*. It is necessary to define a *grey body* as one having constant emissivity at all wavelengths and all temperatures in order to obtain solutions to real problems by treating selective emitters in small increments of wavelength of constant emissivity (Section 5.7). The emissivity of a grey body is

$$\varepsilon = \left[\frac{\dot{E}''_g}{\dot{E}''_b}\right] \quad \text{or} \quad \dot{E}''_g = \varepsilon \dot{E}''_b \quad (2.17)$$

and is illustrated in Figure 2.3. *Kirchoff's* law for grey body radiation states that the emissivity of a grey surface is equal to the absorbtivity of the surface. Thus, for a grey body:

$$\varepsilon = \alpha$$

Energy that is not absorbed by the grey body will be reflected if $\tau = 0$.

2.11 Heat transfer for a grey body in black surroundings

A useful and simple radiation problem is the grey body in black surroundings (Figure 2.4) for which it can be demonstrated that

$$\dot{Q}_{21} = \varepsilon_1 A_1 \sigma (T_2^4 - T_1^4) \quad (2.18)$$

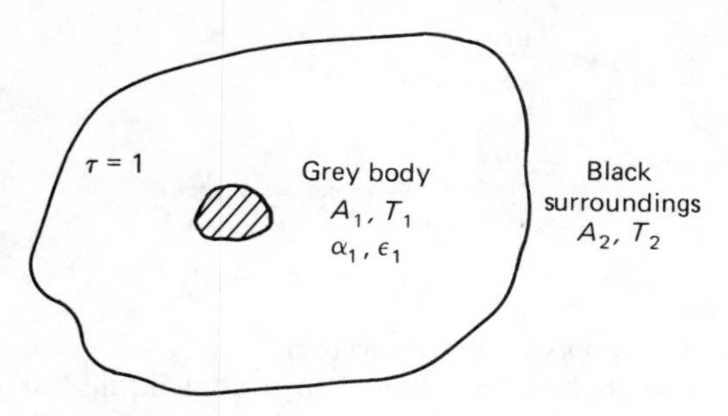

Figure 2.4

where $\dot{Q}_{21}$ is the heat transfer rate from the black surroundings to the grey body, and suffix 1 refers to the grey body and suffix 2 to the black surroundings. This simple equation enables a useful approximation to be made for real bodies, for which a realistic estimate of a constant emissivity may be made, in *large* surroundings. Since the surroundings will not be black the 'large' factor ensures that reflected energy is unlikely to be reincident on the relatively small grey body and will, by successive reflections to the surroundings, become less and less. Any other radiation problems require the work contained in Chapter 5.

2.12 Radiation heat transfer coefficient

It is sometimes convenient to treat radiation by means of a simple radiation heat transfer coefficient based on the temperature difference between radiator and receiver:

$$\dot{Q}''_{\text{radiation}} = h_{\text{radiation}}(T_2 - T_1) = h_{\text{radiation}}\theta \qquad (2.19)$$

For the case of the grey body in large surroundings (by equations (2.18) and (2.19)):

$$h_{\text{radiation}} = \varepsilon_1\sigma(T_1 + T_2)(T_1^2 + T_2^2)$$

This coefficient could be added to the convection surface heat transfer coefficient, enabling simple calculations for heating surfaces which will effect heat transfer by both convection and radiation.

2.13 Simple transient problems in heat transfer

If a body has high thermal conductivity compared with the surface heat transfer coefficient for the particular situation, it is possible to assume that the body temperature is uniform and that in a heating or cooling problem the body temperature is a function of *time* only.

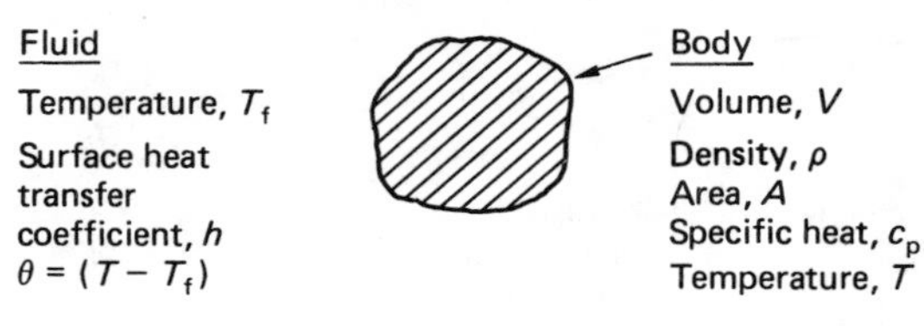

Figure 2.5

Such a body is called a *lumped capacity* system.

An example is a hot mass being quenched in a cool liquid for which Figure 2.5 illustrates the body and nomenclature. In time dt a small heat transfer dQ takes place with a change of body temperature dT:

$$\mathrm{d}Q = -\rho c_p V \mathrm{d}T = hA(T - T_f)\,\mathrm{d}t$$

Hence by integration, putting $(T - T_f) = \theta$ and $\mathrm{d}T = \mathrm{d}\theta$:

$$\frac{\theta}{\theta_0} = \exp\left[-\left(\frac{hAt}{\rho c_p V}\right)\right] \tag{2.20}$$

The elapsed time for the initial temperature difference θ_0 to change to θ is t. More complex transient problems are considered in Chapter 3.

Reference

1. CIBS Guide A3, *Thermal Properties of Building Structures*, London (1980).

WORKED EXAMPLES

Program 2.1 Heat transfer in a plane wall

A plane wall of a building consists of two layers of brick, each 105 mm thick, separated by an air gap with a thermal resistance of 0.18 m^2 K/W. The inside air temperature is 19 °C and the outside air temperature is −5 °C. The inside surface heat transfer coefficient is 5 W/m^2 K and the outside surface heat transfer coefficient is 20 W/m^2 K. (Figure 2.6) The thermal conductivity of brick is 0.84 W/m K. Write a program to find the following:

(a) the inside wall temperature;
(b) the surface temperature on the air gap side of the inside brick;
(c) the heat transfer rate per unit area through the wall.

The three parameters which are to be found are important since:

1. The inside wall temperature affects the comfort of the room.

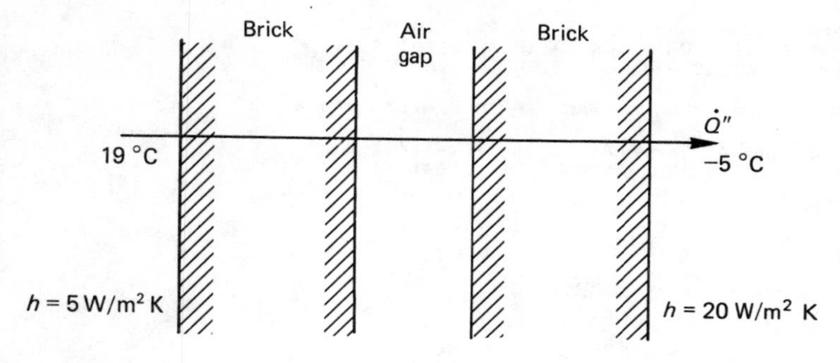

Figure 2.6

2. The air gap wall temperature can cause condensation in the air gap if it is lower than the dew point of the air.
3. The heat transfer shows how much heating is required to maintain the temperature at 19 °C (see Programs 2.2 and 2.3).

The necessary reading for this program is Sections 2.4, 2.5, 2.6 and 2.7.

```
10  PRINT "MULTILAYER PLANE WALL PROBLEM"
20  REM THE PROGRAM CALCULATES HEAT TRANSFER AND INTERFAC
E TEMPERATURES FOR A WALL WITH KNOWN STRUCTURE AND ADJACE
NT FLUID TEMPERATURES
30  PRINT "OUTSIDE AIR TEMPERATURE IS -5 C":TA =  - 5
40  PRINT "INSIDE AIR TEMPERATURE IS 19 C":TI = 19
50  PRINT "OUTSIDE SURFACE HEAT TRANSFER COEFFICIENT IS 2
0 W/M^2 K":HO = 20
60  PRINT "INSIDE SURFACE HEAT TRANSFER COEFFICIENT IS 5
W/M^2 K":HI = 5
70  PRINT "THERMAL RESISTANCE OF AIR GAP IS 0.18 M^2 K/W"
:RG = 0.18
80  PRINT "THICKNESS OF EACH BRICK LAYER IS 0.105 M":X =
0.105
90  PRINT "THERMAL CONDUCTIVITY OF BRICK IS 0.84 W/M K":K
 = 0.84
100  REM  CALCULATE THE TOTAL THERMAL RESISTANCE
110 R = (1 / HO) + (X / K) + RG + (X / K) + (1 / HI)
120 Q = (TI - TA) / R
130  REM  CALCULATE THE INSIDE SURFACE TEMPERATURE AND TH
E SURFACE TEMPERATURE ON THE GAP SIDE OF THE INNER BRICK
140 TSI = TI - (Q / HI)
150 TGI = TSI - (X * Q / K)
160  PRINT
170  PRINT "HEAT TRANSFER THROUGH WALL IS ";Q;" W/M^2"
180  PRINT : PRINT "INSIDE SURFACE TEMPERATURE IS ";TSI;"
 C"
190  PRINT " TEMPERATURE IN THE AIRGAP IS ";TGI;" C"

]RUN
MULTILAYER PLANE WALL PROBLEM
OUTSIDE AIR TEMPERATURE IS -5 C
INSIDE AIR TEMPERATURE IS 19 C
OUTSIDE SURFACE HEAT TRANSFER COEFFICIENT IS 20 W/M^2 K
INSIDE SURFACE HEAT TRANSFER COEFFICIENT IS 5 W/M^2 K
```

```
THERMAL RESISTANCE OF AIR GAP IS 0.18 M^2 K/W
THICKNESS OF EACH BRICK LAYER IS 0.105 M
THERMAL CONDUCTIVITY OF BRICK IS 0.84 W/M K

HEAT TRANSFER THROUGH WALL IS 35.2941177 W/M^2

INSIDE SURFACE TEMPERATURE IS 11.9411765 C
 TEMPERATURE IN THE AIRGAP IS 7.52941177 C
```

Program nomenclature

TA	outside air temperature
TI	inside air temperature
HO	outside surface heat transfer coefficient
HI	inside surface heat transfer coefficient
RG	air gap thermal resistance
X	brick thickness
K	brick thermal conductivity
R	overall thermal resistance
Q	heat transfer rate
TSI	inside surface temperature
TGI	air gap surface temperature

Program notes

(1) The program structure is as follows:

Lines 30–90 set the problem data.
Line 110 calculates the overall thermal resistance.
Line 120 calculates the heat transfer rate.
Lines 140 and 150 calculate temperatures.

(2) Lines 30–90 would be better arranged by asking a question to be displayed on the screen with data supplied in reply, e.g.

30 PRINT "WHAT IS THE OUTSIDE AIR TEMPERATURE?"
40 INPUT TA: PRINT "OUTSIDE AIR TEMPERATURE IS"; TA; "C"

This would enable the program to use different data.

(3) The heat transfer rate is the same through each layer, thus:

(a) to calculate the inside surface temperature we consider convection at this surface:

$$\dot{Q}'' = h\theta$$

(b) to calculate the air gap temperature we consider conduction through the brick:

$$\dot{Q}'' = k\frac{\Delta T}{\Delta x}$$

Program 2.2 Room heater

A room heater has a total area of 2.8 m^2 and a surface temperature of 80 °C. The surface heat transfer coefficient for convection is 11.08 W/m^2 K and the surface emissivity is 0.8. Determine the radiation heat transfer coefficient and add it to the convection heat transfer coefficient to determine the heater output in a room at 19 °C. The Stefan–Boltzmann constant is 5.67 × 10^{-8} W/m^2 K^4. The necessary reading for this program is Sections 2.5, 2.11 and 2.12.

We assume that the grey heater is in large surroundings so tha equation (2.18) may be used. Section 2.12 gives the radiation hea transfer coefficient for this situation as

$$h_{\text{radiation}} = \varepsilon_1\sigma(T_1 + T_2)(T_1^2 + T_2^2)$$

```
10  PRINT "ROOM HEATER"
20  REM  THE PROGRAM CALCULATES THE HEAT TRANSFER BY RADI
ATION AND CONVECTION FRON A FROM HEATER
30  PRINT "HEATER SURFACE TEMPERATURE IS 80 C":TS = 80
40  PRINT "TOTAL SURFACE AREA OF HEATER IS 2.8 M^2":A = 2
.8
50  PRINT "SURFACE HEAT TRANSFER COEFFICIENT IS 11.08 W/M
^2 K":H = 11.08
60  PRINT "EMISSIVITY OF HEATER SURFACE IS 0.8":E = 0.8
70  PRINT "ROOM TEMPERATURE IS 19 C":T = 19
80  REM  STEFAN BOLTZMANN CONSTANT
85 S = 0.0000000567
90  REM  CALCULATE RADIATION HEAT TRANSFER COEFFICIENT TO
 BE ADDED TO CONVECTION HEAT TRANSFER COEFFICIENT
100 HRAD = E * S * ((T + 273) + (TS + 273)) * (((T + 273)
 ^ 2) + ((TS + 273) ^ 2))
110  PRINT "RADIATION HEAT TRANSFER COEFFICIENT IS ";HRAD
;" W/M^2 K"
120  REM  CALCULATE HEAT TRANSFER BY CONVECTION WITH ENHA
NCED COEFFICIENT
130 Q = (H + HRAD) * A * (TS - T)
135  PRINT
140  PRINT "HEAT TRANSFER FROM HEATER IS ";Q;" W"

]RUN
ROOM HEATER
HEATER SURFACE TEMPERATURE IS 80 C
 TOTAL SURFACE AREA OF HEATER IS 2.8 M^2
SURFACE HEAT TRANSFER COEFFICIENT IS 11.08 W M^2 K
EMISSIVITY OF HEATER SURFACE IS 0.8
ROOM TEMPERATURE IS 19 C
RADIATION HEAT TRANSFER COEFFICIENT IS 6.14029635 W/M^2 K

 HEAT TRANSFER FROM HEATER IS 2941.22662 W
```

Program nomenclature

TS	heater surface temperature
A	heater surface area
H	surface heat transfer coefficient for convection
E	emissivity
T	room temperature
S	Stefan–Boltzmann constant
HRAD	radiation heat transfer coefficient
Q	heat transfer rate

Program notes

(1) As in Program 2.1 the data would be better supplied by INPUT statements to allow different problems to be solved.
(2) The program structure is as follows:

Lines 30–85 supply the problem data.
Line 100 calculates radiation heat transfer coefficient.
Line 130 calculates heat transfer rate.

(3) Note that in line 20 *thermodynamic* temperatures are used to calculate radiation heat transfer coefficient but when using the combined coefficient in line 130 they are not. This is because

$$(TS + 273) - (T + 273) = (TS - T)$$

(4) Just under a third of the heat transfer is due to radiation but the size of the convection coefficient suggests that this may be a fan assisted heater. Without fan assistance the convection coefficient would be approximately equal to the radiation coefficient allowing equal contributions from convection and radiation to the heat transfer rate.

Program 2.3 Building heat losses and heaters

In Program 2.1 the heat losses through a plane wall were determined and in Program 2.2 the output from a room heater was calculated. In this program these two pieces of information are brought together to find the number of radiators needed to heat a simple building.

A small village hall of length 12 m, width 6 m and height 5 m is constructed with the walls discussed in Program 2.1 and it may be assumed that the floor and flat roof have the same heat loss per unit area. The building has 30 m^2 of windows with a U value of 3 $W/m^2 K$ and a door area of 2 m^2 with a U value of 0.7 $W/m^2 K$. It is expected that the occupancy will average 20 persons, each of which will give off energy at 120 W. No allowance is made for lighting energy of other small casual grains from, for example, record players or cookers. The outside air temperature is −5 °C and the inside air temperature is 19 °C as in the previous programs. The necessary reading for this program is Section 2.7 and Programs 2.1 and 2.2.

The method calculates the heat losses through the building fabric and subtracts the heat gain from the occupants. The result is divided by the output from one heater to give the number of heaters required.

```
10  PRINT " BUILDING HEATING PROGRAM"
20  REM  THIS PROGRAM CALCULATES HOW MANY RADIATORS OF TH
E TYPE IN THE PREVIOUS PROGRAM ARE NEEDED TO HEAT A SIMPL
E BUILDING WITH WALLS ANALYSED IN THE FIRST PROGRAM
30  REM  ASSUME ROOF AND FLOOR HAVE THE SAME CHARACTERIST
ICS AS THE WALLS,ALLOW FOR A DOOR,WINDOWS AND 20 OCCUPANT
S
40  PRINT "BUILDING DIMENSIONS,HEIGHT 5 M,WIDTH 6 M,LENGT
H 12 M"
50 H = 5:W = 6:L = 12
60  PRINT " TOTAL WINDOW AREA IS 30 M^2":AW = 30
70  PRINT "DOOR AREA IS 2 M^2":AD = 2
80  PRINT "WINDOW U VALUE IS 3 W/M^2 K":UW = 3
90  PRINT " DOOR U VALUE IS 0.7 W/M^2 K":UD = 0.7
100  PRINT " ENERGY EMITTED PER OCCUPANT IS 120 W":EP = 1
20
110  PRINT " FROM PREVIOUS PROGRAM ,HEATER OUTPUT IS 2941
 W":QH = 2941
120  PRINT "FROM PREVIOUS PROGRAM WALL HEAT LOSS IS 35.29
 W/M^2":QW = 35.29
130  PRINT "THERE ARE 20 OCCUPANTS":P = 20
140  PRINT " AIR TEMPERATURES AS IN WALL HEAT LOSS PROGRA
M":TI = 19:TA =  - 5
150  REM  TOTAL EQUIVALENT WALL AREA
160 A = 2 * ((W * L) + (W * H) + (L * H)) - AW - AD
170  REM  TOTAL HEAT LOSS
180 Q = (A * QW) + ((AW * UW) + (AD * UD)) * (TI - TA)
190  REM  HEAT GAIN FROM OCCUPANTS
200 QG = P * EP
210  REM  NUMBER OF RADIATORS
220 N = (Q - QG) / QH
230  PRINT "CALCULATED NUMBER OF RADIATORS IS ";N
240 N =  INT (N) + 1
250  PRINT " NUMBER OF RADIATORS IS ";N
```

```
]RUN
 BUILDING HEATING PROGRAM
BUILDING DIMENSIONS,HEIGHT 5 M,WIDTH 6 M,LENGTH 12 M
 TOTAL WINDOW AREA IS 30 M^2
DOOR AREA IS 2 M^2
WINDOW U VALUE IS 3 W/M^2 K
 DOOR U VALUE IS 0.7 W/M^2 K
 ENERGY EMITTED PER OCCUPANT IS 120 W
 FROM PREVIOUS PROGRAM ,HEATER OUTPUT IS 2941 W
FROM PREVIOUS PROGRAM WALL HEAT LOSS IS 35.29 W/M^2
THERE ARE 20 OCCUPANTS
 AIR TEMPERATURES AS IN WALL HEAT LOSS PROGRAM
CALCULATED NUMBER OF RADIATORS IS 3.43362122
 NUMBER OF RADIATORS IS 4
```

Program nomenclature

H, W, L	building height, width and length
AW	window area
AD	door area
UW	window U value
UD	door U value
EP	energy emitted by a single occupant
QW	wall heat loss per unit area
QH	energy supply from a single heater
P	number of occupants
TI	inside air temperature
TA	outside air temperature
A	total of wall, floor and ceiling areas
Q	total heat loss
QG	total heat gain
N	Number of heaters

Program notes

(1) The program structure is as follows:

Lines 40–140 supply the problem data.
Lines 160–180 calculate heat loss.
Line 200 calculates heat gain.
Lines 210–250 calculate the number of heaters

(2) Again the data would be better given by question and answer using INPUT statements.
(3) Clearly the number of radiators must be a whole number. Line 240 ensures that this is the case.
(4) In problems of this type casual heat gains from people, lighting, cooking, computers, etc., must be included. All power inputs by electricity to a building are dissipated as heat and in summer buildings get hot due to the exposure to radiation from the sun.
(5) The determination of the number of radiators required to maintain the building at 19 °C in an ambient temperature of – 5 °C is probably the worst condition to be expected in Britain. A control system would be required to avoid overheating in milder conditions.

Program 2.4 Economic insulation of a pipe

In general, adding insulation reduces heat loss. With plane walls having constant area this is always true but with tubular or spherical walls the increase in surface area due to addition of insulation can

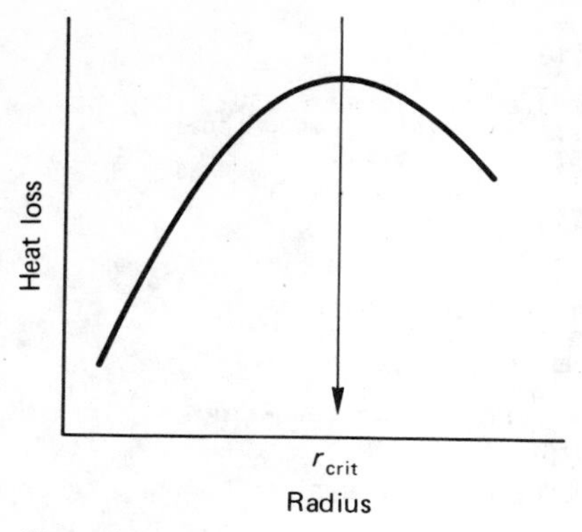

Figure 2.7

in some situations *increase* the heat loss. Analysis for a tube shows that there is a critical radius given by

$$r_{crit} = \frac{k}{h}$$

where k is the thermal conductivity of the insulation and h is the surface heat transfer coefficient. At this radius the heat loss is a maximum (Figure 2.7). If $r < r_{crit}$ the addition of insulation will increase losses. It is usual for the thermal conductivity of insulation to be low (0.03 W/m K) and if h for free convection is of the order 5 W/m^2 K, then critical radius is about 6 mm. This is uncommon for pipes carrying fluids in industry but small pipes and electrical insulation may involve critical radius considerations. For pipes larger than the critical radius, insulation reduces heat transfer and it is possible to determine an economic thickness based on the cost of the heat loss and the cost of the insulation, which must include the installation costs. The payback period obtained can be considered by the user in the light of the availability of capital and the amount he is prepared to invest to reduce future running costs. The necessary reading for this program is Section 2.7.

A metal pipe has an outer diameter of 150 mm and carries fluid at 130 °C. The pipe surface may be assumed to be at 130 °C. The mean ambient temperature is 20 °C. The fluid is heated by fuel of density 850 kg/m^3 costing 15 p per litre, with calorific value 37 000 kJ/kg burned at 80% efficiency. It is proposed to insulate the pipe with material of thermal conductivity 0.04 W/m K and density 160 kg/m^3 which costs 80p per kg plus £20 per metre length to install. In all cases the surface heat transfer coefficient is 8 W/m^2 K. Write a program to enable an economic decision for the amount of insulation.

```
10  PRINT "ECONOMIC INSULATION OF A PIPE"
20  REM  CONSIDER A 1 METRE LENGTH
30  REM   CONSTANT DATA VALUES FROM PROBLEM
40 H = 8:TA = 20:CV = 37000:EF = 0.8:FC = 15:RF = 850
50 DI = 0.150:TI = 130:K = 0.04:IC = 0.8:L = 1:RI = 160
60  REM  BARE PIPE HEAT LOSS COSTS IN PENCE/SEC
70 QB = H * 3.142 * DI * L * (TI - TA)
80 CB = (QB * FC) / (CV * EF * RF)
90  REM  ANNUAL COST IN POUNDS
100 CB = (CB * 60 * 60 * 24 * 365) / 100
105  PRINT
110  PRINT "INSULATION HEAT COST COST OF PAYBACK"
120  PRINT "THICKNESS  LOSS LOSS INSUL,N PERIOD"
130  PRINT "    MM      W/M  POUND POUND   YEARS"
140  PRINT
145 X = ( INT (QB * 10 + 0.5)) / 10:Y = ( INT (CB * 10 +
0.5)) / 10
150  PRINT "  0          ";X;"  ";Y;"      0       -"
160  REM  INSULATION CALCULATION IN 10 MM LAYERS
170  FOR D = 170 TO 250 STEP 10
180  REM  COST OF INSULATION
185 D = D / 1000
190 CI = ((3.142 * (D ^ 2 - DI ^ 2) * RI * IC) / 4) + 20
200  REM  HEAT LOSS;CALCULATE OVERALL THERMAL RESISTANCE
210 RT = (1 / (2 * 3.142 * (D / 2) * H)) + (( LOG (D / DI
)) / (2 * 3.142 * K))
220 Q = (TI - TA) / RT
230 C = (Q * FC) / (CV * EF * RF)
240 C = (C * 60 * 60 * 24 * 365) / 100
250 T = 1000 * (D - DI) / 2
255 D = 1000 * D
260  REM  PAYBACK PERIOD
270 PB = CI / (CB - C)
275 Q = ( INT (Q * 10 + 0.5)) / 10:C = ( INT (C * 10 + 0.
5)) / 10:CI = ( INT (CI * 10 + 0.5)) / 10:PB = ( INT (PB
* 1000 + 0.5)) / 1000:T = ( INT (T * 10 + 0.5)) / 10
280  PRINT  TAB( 3)T; TAB( 11)Q; TAB( 17)C; TAB( 24)CI; T
AB( 30)PB
290  NEXT D

]RUN
ECONOMIC INSULATION OF A PIPE

INSULATION HEAT COST COST OF PAYBACK
THICKNESS  LOSS LOSS INSUL,N PERIOD
   MM       W/M  POUND POUND   YEARS

  0        414.7 78     0       -
  10       150.3 28.3   20.6   .415
  15       116.2 21.9   21     .374
  20       95.7  18     21.4   .356
  25       81.9  15.4   21.8   .348
  30       72    13.5   22.2   .344
  35       64.5  12.1   22.6   .343
  40       58.7  11     23.1   .344
  45       54    10.2   23.5   .347
  50       50.2  9.4    24     .35
```

Program nomenclature

H surface heat transfer coefficient
TA ambient temperature
CV calorific value of fuel
EF efficiency of burning fuel

FC	fuel cost
RF	fuel density
DI	outside pipe diameter, inside insulation diameter
TI	outside pipe temperature, inside insulation temperature
K	thermal conductivity of insulation
IC	insulation cost per kg
L	length of insulated pipe
RI	density of insulation
QB	heat loss from bare pipe
CB	cost of heat loss from bare pipe
X, Y	variables for bare pipe tabulation
D	outer diameter of insulation
CI	total cost of insulation
RT	overall thermal resistance of insulation
Q	heat loss from insulated pipe
C	cost of heat loss from insulated pipe
T	thickness of insulation
PB	payback period

Program notes

(1) The program structure is as follows:

Lines 40 and 50 problem data.
Lines 60–100 bare pipe calculation.
Lines 110–150 table titles and bare pipe results.
Lines 160–280 insulation size, cost, payback calculation loop.

(2) As in earlier programs it would be better to supply data by INPUT methods to allow changes in costs, etc.
(3) In this problem we have considered one year as the period for calculation; this means that the loss cost is in pounds for one year but the insulation cost is fixed and we obtain a payback period of about 4 months. After the 4 months the annual savings will be the difference between the bare pipe costs and the new running costs. Depending on the cash flow situation the numerical minimum payback period may not be the best solution. It may also depend on the pipe length. It is quite clear that a minimum investment in 10 mm of insulation reduces the annual heat loss from £78 to £28.3 and this may be the chosen option.

Program 2.5 Lumped capacity system with a grey body in large surroundings

In Section 2.13 a simple transient problem involving conduction and convection was considered. This program considers a similar radiation and conduction problem in which the effects of convection have been ignored.

A small grey body is heated by radiation in a furnace which may be considered as large surroundings. The emissivity of the body is 0.7 and the surface area is 0.045 m^2. The body has a mass of 2.1 kg, and a specific heat capacity of 850 J/kg K. Write a program to find the time taken for the body temperature to rise from 20°C to 320°C in a furnace whose walls are at 700°C. The Stefan–Boltzmann constant is 5.67×10^{-8} W/m^2 K^4. The program works in 5 K temperature increments and first determines the heat transfer rate by radiation (neglecting convection effects) at any body temperature:

$$\dot{Q} = \varepsilon \sigma A (T_s^4 - T_b^4)$$

where T_s is the furnace wall temperature and T_b the body temperature.

The time taken for the body temperature to rise by 5 K is

$$\frac{5mc_p}{\dot{Q}}$$

where m is the body mass and c_p the body specific heat capacity. The time increments are then summed. The necessary reading for this program is Sections 2.11 and 2.13.

```
10  PRINT "GREY BODY IN LARGE SURROUNDINGS"
20  REM  THE PROGRAM CALCULATES THE TIME TO HEAT A SMALL
GREY BODY BY RADIATION
30  PRINT "THE EMISSIVITY OF THE BODY IS 0.7":E = 0.7
40  PRINT "THE SURFACE AREA OF THE BODY IS 0.045 M^2":A =
 0.045
50  PRINT "THE MASS OF THE BODY IS 2.1 KG":M = 2.1
60  PRINT "THE SPECIFIC HEAT OF THE BODY IS 850 J/KG K":C
P = 850
70  PRINT "INITIAL BODY TEMPERATURE IS 20 C":T1 = 293
80  PRINT "THE FINAL TEMPERATURE OF THE BODY IS 320 C":T2
 = 593
90  PRINT " THE LARGE BLACK SURROUNDINGS(FURNACE) TEMPERA
TURE IS 700 C":TS = 973
100  REM  CALCULATE THE PROCESS IN INTERVALS OF 50 DEGREE
 TEMPERATURE RISES
105  REM  STEFAN BOLTZMANN CONSTANT
110 S = 0.0000000567
120 X = 0
130   FOR TB = 293 TO 593 STEP 5
140 Q = E * S * A * ((TS ^ 4) - (TB ^ 4))
150 DX = (M * CP * 5) / Q
160 X = X + DX
170   NEXT TB
180   PRINT "TIME FOR TEMPERATURE TO REACH 320 C IS ";X;"
SECONDS"

]RUN
GREY BODY IN LARGE SURROUNDINGS
```

```
THE EMISSIVITY OF THE BODY IS 0.7
THE SURFACE AREA OF THE BODY IS 0.045 M^2
THE MASS OF THE BODY IS 2.1 KG
THE SPECIFIC HEAT OF THE BODY IS 850 J/KG K
INITIAL BODY TEMPERATURE IS 20 C
THE FINAL TEMPERATURE OF THE BODY IS 320 C
 THE LARGE BLACK SURROUNDINGS(FURNACE) TEMPERATURE IS 700 C
TIME FOR TEMPERATURE TO REACH 320 C IS 359.814465 SECONDS
```

Program nomenclature

E emissivity of body
A surface area of body
M mass of body
CP specific heat capacity of body
T1 initial body temperature
T2 final body temperature
TS furnace wall temperature
S Stefan–Boltzmann constant
TB temperature of body at any instant
X counter

Program notes

(1) For radiation problems temperatures must be in kelvins.
(2) The program structure is as follows:

Lines 30–90 set the problem data.
Lines 120–160 are the calculation loop.

(3) It is possible to determine the instantaneous radiation heat transfer coefficient and solve this problem by the method of Section 2.13. It would also be easy to account for convection heat transfer with this method. Assume a surface heat transfer coefficient of $5\,W/m^2\,K$ and a fluid temperature of 70 °C and recalculate the time taken to raise the temperature to 320 °C.
(4) The program can be run with different increments in temperature rise to see the variation in results.

PROBLEMS

(2.1) Rearrange Programs 2.1, 2.2 and 2.3 to allow information to be supplied by INPUT in reply to questions displayed on the monitor. This enables the programs to be used with different data.
(2.2) Cavity wall insulation is often added to houses to save fuel costs. The thermal conductivity of a suitable material is 0.046 W/m K. If fuel costs 6p/kW h calculate the annual savings in the village hall of Program 2.3 based on a 30 week, 6 hour heating

need for 3 days per week. Assume the average outside temperature is 1 °C during this period. The cost of cavity wall insulation for this building would be £2250 so that the payback period can also be determined. Compare this approach to savings with the simpler method of reducing the thermostat setting to 18 °C.

(2.3) Write a general program to determine the heating costs for a house or bungalow and hence calculate the costs to heat the home you live in. The program should allow the determination of the effects of loft insulation, cavity wall insulation, double glazing, draught proofing, etc. Compare this with the actual heating bills. If payback periods are determined include some simple allowance for fuel cost inflation. Compare the costs of gas, oil, electric and off-peak electric heating methods. Would it be worth changing the fuel used taking into account any capital costs involved? What effect does climatic variation have? What are 'degree days'?

Data (see also Reference 1 of Chapter 2)

Pitched roof at angle θ made of tiles over roofing felt
$U_r = 1.5\,\mathrm{W/mm^2\,K}$

Pitched roof loft air space resistance	$R_a = 0.16\,\mathrm{m^2\,K/W}$
Uninsulated ceiling resistance	$R_c = 8.2\,\mathrm{m^2\,K/W}$
Thermal conductivity of loft insulation	$k = 0.042\,\mathrm{W/m\,K}$
Thermal conductivity of cavity wall insulation	$k = 0.046\,\mathrm{W/m\,K}$
Overall root resistance	$R = \left(\dfrac{1}{U_r}\right)\cos\theta + R_a + R_c + R_{\mathrm{insulation}}$
Upper storey bare wood floors	$U = 1.7\,\mathrm{W/m^2\,K}$
Carpet $U = 0.06\,\mathrm{W/m^2\,K}$, underlay	$U = 0.045\,\mathrm{W/m^2\,K}$
Suspended wood floors and solid floors (based on the full inside to outside temperature difference)	$U = 0.045\,\mathrm{W/m^2\,K}$
Single glazing $U = 5.6\,\mathrm{W/m^2\,K}$, double glazing	$U = 3.0\,\mathrm{W/m^2}$

Air change rate in an undraughtproofed house 3.5 per hour reduced to 1.9 per hour by draughtproofing

(2.4) The small body of Program 2.5 is to be heated in a gas furnace by convection from the hot gases at 700 °C. In free convection the surface heat transfer coefficient will be $5\,\mathrm{W/m^2\,K}$ but with fan assistance this can be raised to a maximum of $25\,\mathrm{W/m^2\,K}$. Write a program to find the time taken to heat the body from 20 °C to 320 °C for a range of surface heat transfer coefficients. What are the advantages and disadvantages of this method of heating?

(2.5) A chemical plant has 1 km of 150 mm diameter pipe carrying fluid at 25 °C, 0.5 km of 100 mm diameter pipe carrying fluid at 50 °Cand a spherical reactor with a surface area of $125\,\mathrm{m^2}$ at 150 °C. Assume the surface temperatures of these components are equal to those of their contents. The pipes are in the open air at an average temperature of 5 °C and the reactor is in a building at 19 °C. The surface heat transfer coefficient in the building varies from 5 $W/m^2 K$ on a still day to 9 $W/m^2 K$ on a windy day and those outside vary from 10 to 30 $W/m^2 K$ on still and windy days. It is proposed to insulate the plant and three estimates are obtained for pipe insulation 15 mm thick and reactor insulation 25 mm thick:

(a) £30 000 for insulation with thermal conductivity 0.05 W/m K;
(b) £35 000 for insulation with thermal conductivity 0.035 W/m K;
(c) £25 000 for insulation with thermal conductivity 0.06 W/m K.

Assuming that the surface heat transfer coefficients do not change due to insulation and that the distribution of the quoted costs is proportional to the surface areas covered, write a program to enable an economic decision to be made about insulation. Make any assumptions found necessary. The fuel used to maintain the fluids at these temperatures costs 5p/kW h and the plant runs for 24 hours per day for 3 week periods with 1 week for recharging after each run. How would the fluids be maintained at the working temperatures in the pipes?

Open-ended problem

(2.6) Air conditioning requires control of temperature and humidity to maintain comfort. In the UK climate this is necessary in summer when solar loads in well-glazed buildings can be considerable and in winter when heating may make an environment too dry. Make a study of air conditioning techniques and determine the maximum loads on a building such as a hotel, office block, factory complex, etc., in summer and in winter. Load will include gains or losses through the fabric, gains in solar energy through windows and by transient effects from the fabric, gains from occupants, lighting, cooking, machinery, etc. Air changes can cause gains or losses [1]. Finally write a simple program to determine these loads and hence, using plant sizing charts [2], determine the space needed for plant and the maximum air conditioning duct size required [3]. (A real local building might be a useful study.)

References

1. CIBS Guide, Sections A3–A9.
2. Burberry, P. *Environment and Services*, Halstead Press (1979).
3. CIBS Guide, Section C4.

Chapter 3

Conduction

3.1 Introduction

In Chapter 2, simple, one-dimensional, steady state conduction situations were considered. In order to proceed to the solution of two- or three-dimensional problems in steady state and transient conditions, further analysis is required to develop a general equation for conduction. Solution of this equation is necessary to determine, for example, the temperature distributions in the cylinder block or pistons of a motor car engine.

3.2 General conduction equation

By considering the thermal equilibrium of a small element of solid material (Figure 3.1(a)) it is possible to draw up a balance between

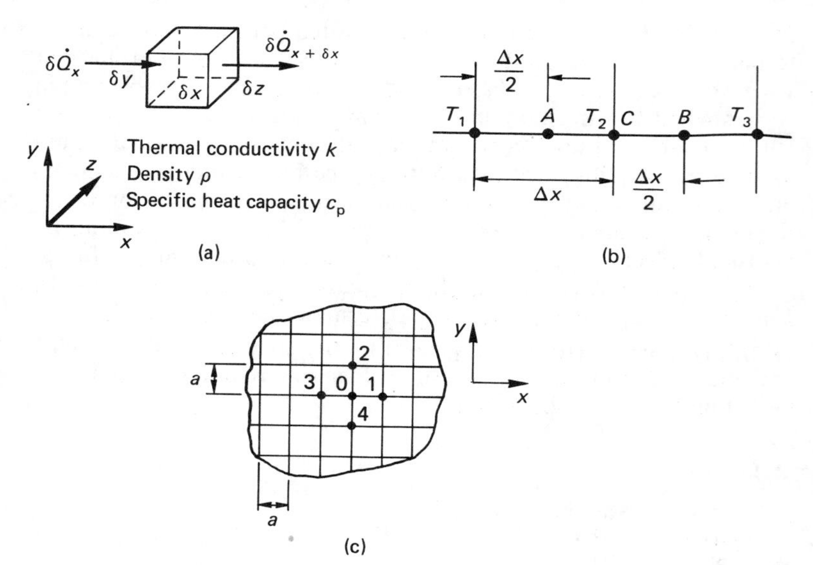

Figure 3.1

the conduction heat transfer to and from the element, the energy stored due to change in element temperature with time and energy generation within the element due, most often,to the conduction of electricity. In rectangular coordinates (x, y, z) the general equation is

$$\frac{\partial T}{\partial t} = \alpha\left(\frac{\partial^2 T}{\partial x^2} + \frac{\partial^2 T}{\partial y^2} + \frac{\partial^2 T}{\partial z^2}\right) + \frac{\dot{Q}'''}{lc_{\mathrm{p}}} \qquad (3.1)$$

where $\frac{\partial T}{\partial t}$ is the rate of change of temperature with time

$\dot{Q}'''$ is the rate of energy generation per unit volume

and α is the thermal diffusivity, $\alpha = \frac{k}{\rho c_{\mathrm{p}}}$

This relation may also be expressed in cylindrical coordinates (r, z, θ):

$$\frac{\partial T}{\partial t} = \alpha\left[\frac{1}{r}\frac{\partial T}{\partial r} + \frac{\partial^2 T}{\partial r^2} + \frac{\partial^2 T}{\partial z^2} + \frac{1}{r^2}\left(\frac{\partial^2 T}{\partial \theta^2}\right)\right] + \frac{\dot{Q}'''}{\rho c_{\mathrm{p}}} \qquad (3.2)$$

Solutions to equation (3.1) will be considered in two-dimensional steady state and in one- and two-dimensional (program 3.5) transient situations.

3.3 Finite difference formulation

Exact solutions to equations (3.1) and (3.2) are only possible in a limited number of cases. Numerical solution by discretisation techniques such as finite difference or finite element are now widely used in software. In this chapter, solution by *finite difference* numerical approximation is used in which continuously varying temperatures are considered to occur in small finite steps of space and time. Consider the three planes Δx apart at which the temperatures are T_1, T_2, T_3 (Figure 3.1(b)).

At A

$$\frac{\partial T}{\partial x} = \left(\frac{T_1 - T_2}{\Delta x}\right)$$

and at B

$$\frac{\partial T}{\partial x} = \left(\frac{T_2 - T_3}{\Delta x}\right)$$

Thus at C:

$$\frac{\partial^2 T}{\partial x^2} = \frac{\left[\left(\dfrac{T_1 - T_2}{\Delta x}\right) - \left(\dfrac{T_2 - T_3}{\Delta x}\right)\right]}{\Delta x} = \frac{T_1 + T_3 - 2T_2}{\Delta x^2} \tag{3.3}$$

Similarly for $\partial T/\partial t$ we consider a finite timestep Δt and let suffix n represent the nth plane, suffix 0 the initial time, and suffix 1 the time after the step Δt:

$$\frac{\partial T}{\partial t} = \frac{T_{n,1} - T_{n,0}}{\Delta t} \tag{3.4}$$

3.4 Steady state two-dimensional conduction

The equation to be solved is

$$\alpha\left(\frac{\partial^2 T}{\partial x^2} + \frac{\partial^2 T}{\partial y^2}\right) + \frac{\dot{Q}'''}{\rho c_\mathrm{p}} = 0$$

and the finite difference technique is applied to the grid shown in Figure 3.1(c) for which we write

$$\alpha\left[\left(\frac{T_1 + T_3 - 2T_0}{a^2}\right) + \left(\frac{T_2 + T_4 - 2T_0}{a^2}\right)\right] + \frac{\dot{Q}'''}{\rho c_\mathrm{p}} = 0$$

With *no heat generation* $\dot{Q}''' = 0$ and we obtain

$$T_1 + T_2 + T_3 + T_4 - 4T_0 = 0 \tag{3.5}$$

With *heat generation* due to electrical conduction with a current density i (amp/m^2) and resistivity r (ohm m)

$$\dot{Q}''' = i^2 r$$

and

$$T_1 + T_2 + T_3 + T_4 - 4T_0 = -\frac{a^2 i^2 r}{k} \tag{3.6}$$

Equation (3.5) or (3.6) must be satisfied for each grid point. Sets of equation (3.5) or (3.6) can be solved by first estimating the temperatures and then by iterative application of equation (3.5) or (3.6) in sequence – updating the temperatures at the end of each sequence (Jacobi iteration), or by iteration in sequence – updating the temperatures as each new value is obtained (Gauss–Seidel iteration). The latter converges on a solution more quickly and there are accelerated convergence techniques [1]. Sets of equations may also be solved by Gaussian elimination to give an exact solution to the approximation.

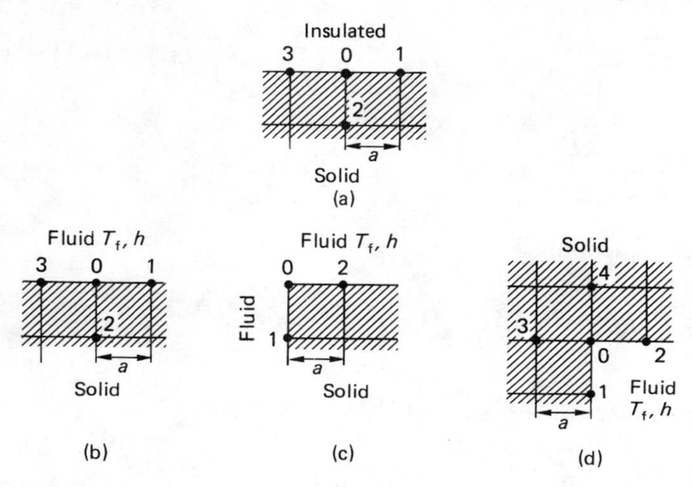

Figure 3.2

3.5 Boundary conditions

Equation (3.5) applies to interior grid points of the solid body. At the boundaries different equations will apply depending on the boundary conditions. Common conditions include the following

(a) isothermal boundary for which no equation has to be solved;
(b) insulated boundary with no heat generation.

Draw up an energy balance at the boundary (Figure 3.2(a)). For unit height:

$$k \cdot a \cdot 1 \left(\frac{T_2 - T_0}{a} \right) + k \cdot \frac{a}{2} \cdot 1 \left(\frac{T_1 - T_0}{a} \right) + k \cdot \frac{a}{2} \cdot 1 \left(\frac{T_3 - T_0}{a} \right)$$

$$= 0 \times T_1 + T_3 + 2T_2 - 4T_0 = 0 \qquad (3.7)$$

With heat generation:

$$T_1 + T_3 + 2T_2 - 4T_0 = -\frac{a^2 i^2 r}{2k} \qquad (3.8)$$

(c) Convective boundary with no heat generation.

Let the fluid temperature be T_f and the surface heat transfer coefficient be h and draw up an energy balance at the boundary (Figure 3.2(b)). For unit height:

$$k \cdot a \cdot l \left(\frac{T_2 - T_0}{a} \right) + k \cdot \frac{a}{2} \cdot l \left(\frac{T_1 - T_0}{a} \right) + k \cdot \frac{a}{2} \cdot l \left(\frac{T_3 - T_0}{a} \right)$$

$$+ h \cdot a \cdot l (T_f - T_0) = 0$$

$$\frac{T_1}{2} + \frac{T_3}{2} + T_2 + \frac{ha}{k} \cdot T_f - T_0 \left(2 + \frac{ha}{k} \right) = 0 \tag{3.9}$$

With heat generation:

$$\frac{T_1}{2} + \frac{T_3}{2} + T_2 + \frac{ha}{k} \cdot T_f - T_0 \left(2 + \frac{ha}{k} \right) = -\frac{a^2 i^2 r}{2k} \tag{3.10}$$

External corner (Figure 3.2(c)) by similar methods:

$$\frac{T_2}{2} + \frac{T_1}{2} + \frac{ha}{k} \cdot T_f - T_0 \left(1 + \frac{ha}{k} \right) = 0 \quad \text{or} \quad -\frac{a^2 i^2 r}{4k} \tag{3.11}$$

Internal corner (Figure 3.2(d)) by similar methods:

$$\frac{T_1}{2} + \frac{T_2}{2} + T_3 + T_4 + \frac{ha}{k} \cdot T_f - T_0 \left(3 + \frac{ha}{k} \right) = 0$$

$$\text{or} \quad -\frac{3a^2 i^2 r}{4k} \tag{3.12}$$

Equations for other boundary conditions and shapes are available [1].

3.6 Heat transfer rates at a boundary

Conceptually the grid lines may be imagined as conducting rods of unit height and width a with thermal conductivity k situated symmetrically on the grid (Figure 3.3). This enables heat transfer rates at

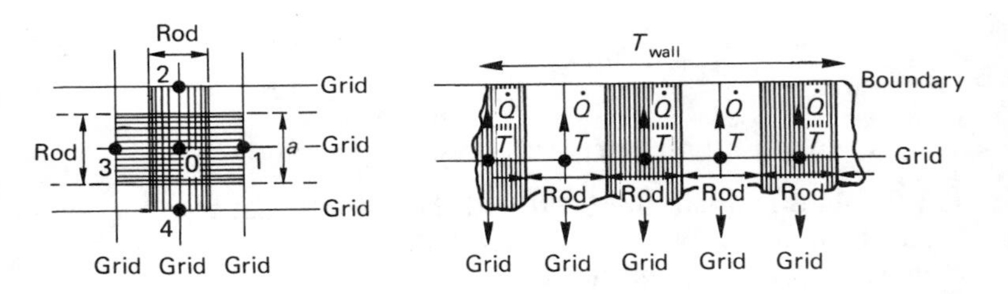

Figure 3.3

the boundary to be determined. The solution to equations (3.5)–(3.12) will give a temperature distribution. By the rod analogy the heat transfer rate per unit height to an isothermal boundary is given by

$$\dot{Q}' = \sum k(T - T_{\text{wall}}) \tag{3.13}$$

where T is the temperature of each grid point which is connected to the boundary by a conducting rod (grid line).

At a *convective boundary* the heat transfer rate per unit height to the boundary plane is given by

$$\dot{Q}' = \sum ha(T_{\text{f}} - T) \tag{3.14}$$

where T is the temperature at each grid point on the boundary.

3.7 One-dimensional transient conduction

The equation to be solved is

$$\frac{\partial T}{\partial t} = \alpha \frac{\partial^2 T}{\partial x^2} + \frac{\dot{Q}'''}{\rho c_{\text{p}}}$$

and the finite difference technique is applied to the nth plane (Figure 3.4(a)) for the case when $\dot{Q}'''$ is zero. It should be noted that each temperature has two suffixes: the plane n, $n + 1$, $n - 1$ and the time 0 or 1.

$$\frac{T_{n,1} - T_{n,0}}{\Delta t} = \alpha \left[\frac{T_{n-1,0} + T_{n+1,0} - 2T_{n,0}}{a^2} \right]$$

$$T_{n,1} = \frac{\alpha \Delta t}{a^2} [T_{n-1,0} + T_{n+1,0} - 2T_{n,0}] + T_{n,0}$$

Let $\alpha \Delta t / a^2 = F$, the (non-dimensional) grid size Fourier number:

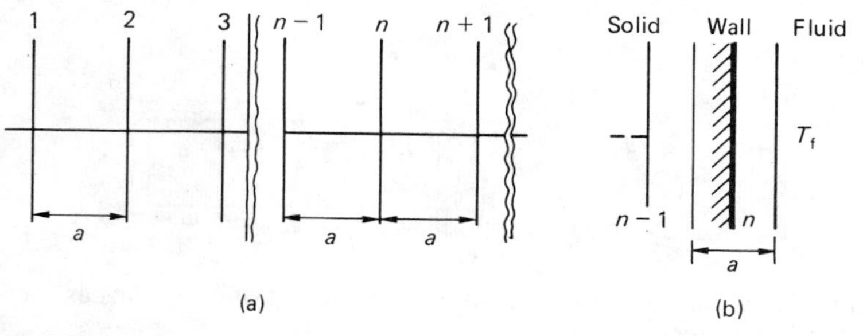

Figure 3.4

$$T_{n,1} = F\left(T_{n-1,0} + T_{n+1,0} + T_{n,0}\left(\frac{1}{F} - 2\right)\right) \quad (3.15)$$

Equation (3.15) has been formulated with *backward* time differences to the known temperatures at time 0 to produce *explicitly* the temperatures at time 1 at each plane in succession. Although this gives an easy solution it has limitations because examination of equation (3.15) shows that if $F > 0.5$ the solution becoms oscillatory and unstable. It is also found that temperatures are underestimated. The explicit method is also known as the Euler method. The problems of stability can be solved by using *forward* time differences and writing

$$\frac{T_{n,1} - T_{n,0}}{\Delta t} = \alpha\left[\frac{T_{n-1,1} + T_{n+1,1} - 2T_{n,1}}{a^2}\right]$$

giving

$$-F \cdot T_{n-1,1} + (1 + 2F)T_{n,1} - F \cdot T_{n+1,1} = T_{n,0} \quad (3.16)$$

In equation (3.16) the only known temperature is $T_{n,0}$ and it cannot be solved alone. A set of equations for each plane can be solved by Gaussian elimination to yield all the temperatures at time 1. This is known as the implicit method and it is stable for all F. Temperatures are overestimated by this method. To avoid problems of over- and underestimates a method based on the average of backward and forward time differences was developed by Crank and Nicolson which gives good results with some stable oscillations at high F [2]. In this formulation

$$\frac{T_{n,1} - T_{n,0}}{\Delta t} = \frac{\alpha}{2} \times \left[\frac{T_{n+1,0} + T_{n-1,0} - 2T_{n,0} + T_{n+1,1} + T_{n-1,1} - 2T_{n,1}}{a^2}\right]$$

giving

$$-\frac{F}{2} \cdot T_{n-1,1} + (1 + F)T_{n,1} - \frac{F}{2} \cdot T_{n+1,1} = \frac{F}{2} \cdot T_{n-1,0} + (1 + F)T_{n,0} + \frac{F}{2} \cdot T_{n+1,0}$$

Gaussian elimination will be required for the set of equations as the method is implicit.

3.8 Boundary conditions

Boundary conditions in transient problems are likely to include sudden rises in temperature to an isothermal state or a steady rise in temperature with time. Convective boundaries are common and periodic variations in temperature in convective situations are found. For convective boundaries an energy balance at the surface must be made (Figure 3.4(b)).

Let the surface area be **A** and using the backward (explicit) time difference format:

$$-k\cdot A\left(\frac{T_{n,0} - T_{n-1,0}}{a}\right)\Delta t - h\cdot A(T_{n,0} - T_{\mathrm{f}})\Delta t$$

$$= \rho\left(\frac{a}{2}\cdot A\right)c_{\mathrm{p}}(T_{n,1} - T_{n,0})$$

Put $\alpha\Delta t/a^2 = \mathrm{F}$ (Fourier number) and $ha/k = B$ the (non-dimensional) grid Biot number; then

$$T_{n,1} = 2F\cdot T_{n-1,0} + T_{n,0}(1 - 2F - 2F\cdot B) + 2F\cdot B\cdot T_{\mathrm{f}} \quad (3.18)$$

Equation (3.18) will be oscillatory if $F + FB > 0.5$ which is a more stringent limitation than the body plane condition. Equation (3.18) can be formulated in forward and average time differences for the implicit methods.

3.9 Significance of the Fourier number, *F*

The Fourier number is important because it involves the grid size and the timestep. If it is limited to 0.5 (one-dimensional isothermal boundary case) and a grid size is chosen, then the time interval is fixed and it may well be inconveniently long or short for the information required. An implicit method may be better suited. If F is chosen large for an implicit method, then the timesteps may be too large to give accurate answers; F must be carefully selected to obtain sensible results. The stability criteria to the explicit method becomes more severe in two and three dimensions (in two dimensions with isothermal boundaries $F > 0.25$) and again an implicit method may be chosen. Clearly, the Crank–Nicolson method, which gives more accurate temperatures, will often be the most satisfactory route with reasonable freedom of choice for F.

3.10 Heat transfer

The results of the solution of a transient equation give a series of

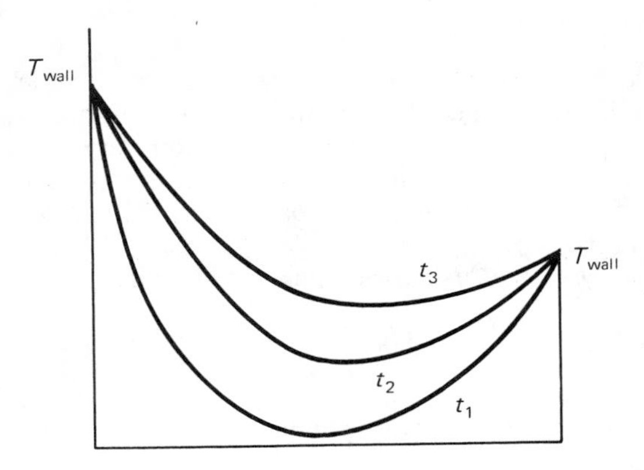

Figure 3.5

temperatures within a body at a number of points at particular times. Temperature profiles can be obtained for various times, and interpolation by hand or computer will give information about other points at intermediate times (Figure 3.5). Instantaneous heat transfer rates (watts) to the body can be calculated from

$$\dot{Q} = kA\left(\frac{\Delta T}{\Delta x}\right)_{\text{boundary at time } t}$$

but it is more likely that the heat transfer to achieve a given temperature or the time taken to reach a given temperature are required. The latter is obtained by the interpolation described above. The heat transfer (joules) is obtained from either the sum of the instantaneous heat transfer rates multiplied by the timestep:

$$Q = \sum_{\text{all time}} kA\left(\frac{\Delta T}{\Delta x}\right)_{\substack{\text{boundary}\\ \text{at time } t}} \cdot \Delta t \qquad (3.19)$$

or from the temperature profile at any time:

$$Q = \sum_{\text{all layers}} mc_{\text{p}}(T_{\text{final time}} - T_{\text{zero time}}) \qquad (3.20)$$

Equations (3.19) and (3.20) are for one-dimensional problems.

3.11 Transient heat transfer charts

Because of the historical necessity to solve problems, charts are available for common shapes (slabs, cylinders, spheres) which give transient temperatures and heat transfers. Methods exist for using the charts to produce results for other solids [3]. These charts are useful for checking programs, as are the available analytical solutions.

References

1. Myers. G.E. *Analytical Methods in Conduction Heat Transfer*, McGraw-Hill (1971).
2. *Proceedings of the Cambridge Philosophical Society*, **43**, 50–67 (1947).
3. White, F.M. *Heat Transfer*, Addison-Wesley (1984).

WORKED EXAMPLES

Program 3.1 Two-dimensional steady state conduction

Figure 3.6 shows a chimney of rectangular cross-section with isothermal walls. Write a program to determine the temperature distribution in the chimney and the heat transfer rate to and from the chimney, obtaining solutions by (i) an iterative method and (ii) Gaussian elimination.

Data

Chimney dimensions	2 m × 2 m outside
	0.8 m × 0.8 m inside
Thermal conductivity of material	0.9 W/m K

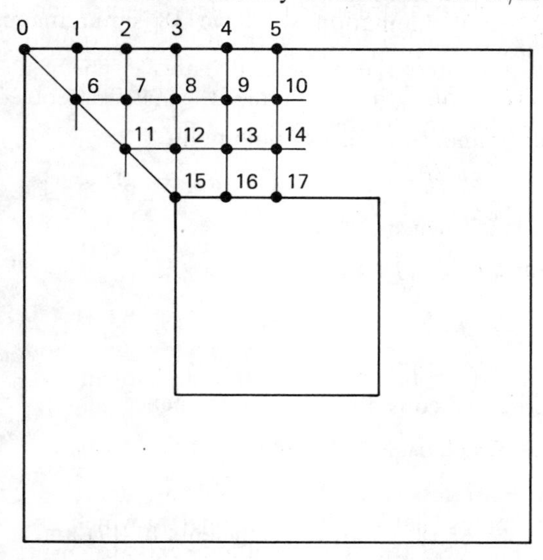

Figure 3.6

Temperature outer wall 20 °C
inner wall 255 °C

The necessary reading for this program is Sections 3.4, 3.5 and 3.6.

A mesh 0.2 m square is chosen and examination of the symmetry shows that only one-eighth need be considered. The equation to be satisfied at the body points is equation (3.5) so that at body point 8, for example, we have

$$T_3 + T_7 + T_9 + T_{12} - 4T_8 = 0$$

and at body point 10, we have

$$T_5 + 2T_9 + T_{14} - 4T_{10} = 0$$

Other points are similar.

1. To solve by iteration we assign arbitrary values to all the unknowns (100 °C in this problem). In the solution shown, Jacobi iteration is used and the temperatures are updated after each complete iteration. After three iterations the results are displayed. However, it is possible to allow the program to decide when solution is completed and this is done by testing to see if the change in the temperatures between the current iteration and the previous iteration is less than 0.5 °C. If this is so iteration ceases, the last being printed, and the heat transfers are calculated. A Gauss–Seidel iteration would update each temperature as it is found (Section 3.4).
2. To solve by Gaussian elimination we have 18 simultaneous equations, each of which will appear as

$$a_0 T_0 + a_1 T_1 + a_2 T_2 + \ldots a_{17} T_{17} = \text{constant}$$

At the isothermal boundaries, for example point 1:

$$a_0 = 0, \quad a_1 = 1, \quad a_2 \text{ to } a_{17} = 0$$

and the constant is 25; hence

$$0T_0 + 1T_1 + 0T_2 + \ldots 0T_{17} = 25$$

At body point 8:

$$a_3 = 1, a_7 = 1, a_9 = 1, a_{12} = 1 \text{ and } a_8 = -4$$

the rest being zero. The constant is also zero, hence

$$0T_0 \ldots + 1T_3 + 0T_4 \ldots + 1T_7 - 4T_8 + 1T_9 + 0T_{10} \ldots$$
$$+ 1T_{12} + 0T_{13} \ldots = 0$$

Thus a matrix of 18 lines of 18 equation coefficients (0–17) and a constant can be written as data for use with the Gaussian elimination Program 1.2. These 18 lines are shown in the listing lines 520–690.

```
10  PRINT "STEADY TWO DIMENSIONAL CONDUCTION "
15  PRINT "WITH ISOTHERMAL BOUNDARY CONDITIONS"
20  REM IN THIS PROBLEM THERE ARE 18 NODES STATEMENT 30 M
UST ALLOW ENOUGH SPACE
30  DIM T(40): DIM B(40)
40  FOR N = 0 TO 17
50  READ T(N)
55 B(N) = T(N)
60  NEXT N
70  REM  THE 9 POINTS WITH UNKNOWN TEMPERATURES ARE NUMBE
RS 6 TO 14
80  REM  POINTS 7,8 AND 9 ARE SIMILAR;POINTS 12 AND 13 AR
E SIMILAR AND POINTS 6,10,11 AND 14 ARE SINGULAR SO THAT
SIX DIFFERENT EQUATIONS ARE INVOLVED
90  PRINT
100 A = 0
110 N = 6
120 B(N) = (2 * T(1) + 2 * T(7)) / 4
130  FOR N = 7 TO 9
140 B(N) = (T(N - 5) + T(N + 4) + T(N + 1) + T(N - 1)) /
4
150  NEXT N
160 N = 10
170 B(N) = (T(5) + T(14) + 2 * T(9)) / 4
180 N = 11
190 B(N) = (2 * T(7) + 2 * T(12)) / 4
200  FOR N = 12 TO 13
210 B(N) = (T(N - 4) + T(N + 3) + T(N + 1) + T(N - 1)) /
4
220  NEXT N
230 N = 14
240 B(N) = (T(10) + T(17) + 2 * T(13)) / 4
250 A = A + 1
270  REM  ALLOW THREE ITERATIONS AND INSPECT THE RESULT
280  IF A = 3 THEN  GOTO 580
290  IF A < 3 THEN  GOTO 350
300  REM  FURTHER ITERATIONS COULD BE PRINTED UNTIL SATIS
FIED BUT THE COMPUTER CAN DETERMINE SATISFACTION
310  FOR N = 6 TO 14
320  IF  ABS (B(N) - T(N)) > 0.5 THEN  GOTO 350
330  NEXT N
340  GOTO 400
350  FOR N = 6 TO 14
360 T(N) = B(N)
370  NEXT N
380  GOTO 110
400  PRINT
410  PRINT "TEMPERATURES AFTER ";A;" ITERATIONS"
420  PRINT
430  PRINT "POINT          TEMPERATURE"
440  PRINT
450  FOR N = 6 TO 14
460  PRINT  TAB( 3)N, TAB( 10)T(N)
470  NEXT N
480  REM  DETERMINE THE HEAT TRANSFER RATES
490 QIN = (4 * (225 - T(14)) + 8 * (225 - T(13)) + 8 * (2
25 - T(12))) * 0.9
500 QOT = (4 * (T(10) - 25) + 8 * ((T(9) - 25) + (T(8) -
25) + (T(7) - 25) + (T(6) - 25))) * 0.9
510  PRINT
520  PRINT "HEAT TRANSFER RATE TO CHIMNEY IS ";QIN;" W/M"
```

```
530  PRINT "HEAT TRANSFER FROM CHIMNEY IS ";QOT;" W/M"
540  END
580  PRINT
590  PRINT "TEMPERATURE AFTER THREE ITERATIONS"
595  PRINT
600  PRINT "POINT          TEMPERATURE"
610  PRINT
620  FOR N = 6 TO 14
630 T(N) = B(N)
640  PRINT  TAB( 3)N, TAB( 10)T(N)
650  NEXT N
660  GOTO 110
1000  DATA  25,25,25,25,25,25,100,100,100,100,100,100,100
,100,100,225,225,225

]RUN
STEADY TWO DIMENSIONAL CONDUCTION
WITH ISOTHERMAL BOUNDARY CONDITIONS


TEMPERATURE AFTER THREE ITERATIONS

POINT            TEMPERATURE

  6              46.09375
  7              66.015625
  8              76.5625
  9              81.640625
  10             81.640625
  11             100.78125
  12             138.28125
  13             145.3125
  14             147.265625

TEMPERATURES AFTER 10 ITERATIONS

POINT          TEMPERATURE

  6              43.013382
  7              61.3706589
  8              76.9787788
  9              84.7249031
  10             86.5002632
  11             99.3160248
  12             138.041306
  13             149.830818
  14             152.902699

HEAT TRANSFER RATE TO CHIMNEY IS 1425.871 W/M
HEAT TRANSFER FROM CHIMNEY IS 1417.23255 W/M


520  DATA  1,0,0,0,0,0,0,0,0,0,0,0,0,0,0,0,0,0,25
530  DATA  0,1,0,0,0,0,0,0,0,0,0,0,0,0,0,0,0,0,25
540  DATA  0,0,1,0,0,0,0,0,0,0,0,0,0,0,0,0,0,0,25
550  DATA  0,0,0,1,0,0,0,0,0,0,0,0,0,0,0,0,0,0,25
560  DATA  0,0,0,0,1,0,0,0,0,0,0,0,0,0,0,0,0,0,25
570  DATA  0,0,0,0,0,1,0,0,0,0,0,0,0,0,0,0,0,0,25
580  DATA  0,2,0,0,0,0,-4,2,0,0,0,0,0,0,0,0,0,0,0
590  DATA  0,0,1,0,0,0,1,-4,1,0,0,1,0,0,0,0,0,0,0
600  DATA  0,0,0,1,0,0,0,1,-4,1,0,0,1,0,0,0,0,0,0
610  DATA  0,0,0,0,1,0,0,0,1,-4,1,0,0,1,0,0,0,0,0
620  DATA  0,0,0,0,0,1,0,0,0,2,-4,0,0,0,1,0,0,0,0
630  DATA  0,0,0,0,0,0,0,2,0,0,0,-4,2,0,0,0,0,0,0
640  DATA  0,0,0,0,0,0,0,0,1,0,0,1,-4,1,0,1,0,0,0
650  DATA  0,0,0,0,0,0,0,0,0,1,0,0,1,-4,1,0,1,0,0
660  DATA  0,0,0,0,0,0,0,0,0,0,1,0,0,2,-4,0,0,1,0
670  DATA  0,0,0,0,0,0,0,0,0,0,0,0,0,0,0,1,0,0,225
```

```
680  DATA  0,0,0,0,0,0,0,0,0,0,0,0,0,0,0,0,1,0,225
690  DATA  0,0,0,0,0,0,0,0,0,0,0,0,0,0,0,0,0,1,225

]RUN
GAUSSIAN ELIMINATION PROGRAM

NUMBER OF EQUATIONS?18
 SOLVING 18 EQUATIONS

SOLUTION IS
X1= 25
X2= 25
X3= 25
X4= 25
X5= 25
X6= 25
X7= 43.14
X8= 61.28
X9= 77.31
X10= 84.89
X11= 86.98
X12= 99.67
X13= 138.06
X14= 150.27
X15= 153.13
X16= 225
X17= 225
X18= 225
```

Program nomenclature

T, B temperatures at the mesh points
A iteration counter
QIN heat transfer rate per unit height from the flue gas wall
QOT heat transfer rate per unit height to the air wall.

Program notes

(1) The program structure is as follows:

Lines 30–50 give dimension space and read in T data.
Line 35 sets up a dummy variable B for temperature.
Lines 70–240 complete an iteration of B.
Line 250 counts the iterations.
Lines 220–320 display the third iteration with lines 600 to 650 and then allow iteration to continue till the required accuracy is achieved when the results are printed and lines 480–540 calculate the heat transfer rates. If the required level of accuracy has not been reached the temperatures are updated by lines 350–380 and the next iteration starts.

(2) The results show that after three iterations the temperatures range between 46 °C and 147 °C and after 10 iterations no change greater than 0.5 °C has occurred from the ninth iteration. The temperatures range from 43 °C to 153 °C. These are *approximate*

answers to the finite difference *approximate* equations. The Gaussian elimination results which follow are *exact* solutions to the *approximate* equations.

(3) What might happen if in line 320 the ABS function was omitted?

IF (B(N) − T(N)) > 0.5 THEN GOTO 350

Program 3.2 One-dimensional transient conduction, explicit method

This program solves the problem where a slab of material at uniform temperature suddenly has the wall temperatures raised to a new isothermal value. In such a situation the body temperature will gradually rise as the energy is conducted from the walls towards the centre. Ultimately, the body would reach a new steady state. In three dimensions this could be the heat treatment of a body, a problem of cookery or fish freezing, and the solution will give the time for a particular point to reach a particular temperature, or the temperature reached at a point after a particular time, together with the ability to calculate the associated heat transfer required. Figure 3.7(a) shows plots of temperature throughout a slab at times 1, 2, 3, 4, etc. The necessary reading for this program is Sections 3.7, 3.9 and 3.10.

This method, using equation (3.15), uses backward time differences and the solution for temperature is explicit:

$$T_{n,1} = F\left(T_{n-1,0} + T_{n+1,0} + T_{n,0}\left[\frac{1}{F} - 2\right]\right)$$

However, for stability of solution:

$$F = \frac{\alpha \Delta t}{a^2} \leqslant 0.5$$

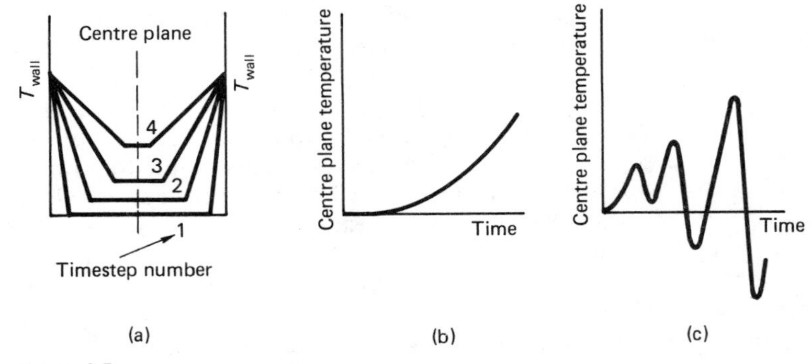

Figure 3.7

Figure 3.7(b) shows the centre line slab temperature with $F < 0.5$ (Run 1). Figure 3.7(c) shows the centre line temperature with $F > 0.5$ (Run 2).

The program considers an elapsed time of 10 minutes and displays the results at one chosen plane at each time step followed by the temperature distribution after 10 minutes.

The problem sets the slab at 0 °C initially and the walls are raised to 100 °C at time zero. The thermal diffusivity of the slab material is $2 \times 10^{-6}\,\mathrm{m^2/s}$, the thermal conductivity is 2 W/m K and the total slab thickness is 0.1 m.

```
10  PRINT "ONE DIMENSIONAL TRANSIENT CONDUCTION"
20  PRINT "ISOTHERMAL BOUNDARY"
30  REM  CHOOSE A NUMBER OF LAYERS NOT EXCEEDING 10 ;10 L
AYERS HAVE 11 PLANES
40  PRINT "HOW MANY LAYERS ARE CHOSEN?"
50  INPUT N: PRINT "THERE ARE ";N;" LAYERS"
60  PRINT "AT WHICH PLANE DO YOU WISH TO "
70  PRINT "DISPLAY THE TEMPERATURE?"
80  INPUT E: PRINT "PLANE ";E
90 W = 0.1 / N
100  REM  A IS THE THERMAL DIFFUSIVITY IN M^2/S AND K THE
 THERMAL CONDUCTIVITY IN W/M K
110 A = 0.000002:K = 2
120  PRINT "CHOOSE THE FOURIER NUMBER"
125  REM  FOR STABILITY F<=0.5
130  INPUT F: PRINT "FOURIER NUMBER IS ";F
140  REM  DETERMINE THE TIME STEP
150 D = F * (W ^ 2) / A
160  PRINT " CONSIDER AN ELAPSED TIME OF 10 MINUTES"
170  PRINT "TIME STEP IS ";D;" SECONDS"
180 B = 600 / D:B = ( INT (B + 0.5))
190  PRINT "NUMBER OF TIME STEPS ";B
200  DIM T(11,B)
210  REM   THE WALLS ARE SUDDENLY RAISED IN TEMPERATURE F
ROM 0 C TO 100 C
220  PRINT
230  FOR X = 1 TO N + 1
240 T(X,0) = 0
250  NEXT X
260  FOR S = 1 TO B: FOR X = 2 TO N
270 T(1,S) = 100:T(N + 1,S) = 100
280 T(X,S) = F * (T(X + 1,S - 1) + T(X - 1,S - 1) + T(X,S
 - 1) * ((1 / F) - 2))
290 T(X,S) = ( INT (T(X,S) * 10 + 0.5)) / 10
300  NEXT X
310  PRINT "TEMPERATURE AT PLANE ";E;" IS ";T(E,S)
320  NEXT S
323  PRINT
324  PRINT "TEMPERATURE DISTRIBUTION AFTER 10 MINUTES"
325  FOR X = 1 TO N + 1: PRINT "PLANE ";X;" T= ";T(X,B);"
 C"
326  NEXT X
327  PRINT
330 M = T(1,B) + T(N + 1,B)
340  FOR X = 2 TO N
350 M = M + (T(X,B) * 2)
360  NEXT X
370 Q = (W * K * M) / (2 * A)
380 Q = ( INT (Q * 10 + 0.5)) / 10
390  PRINT "HEAT TRANSFER IN 10 MINUTES IS ";Q;" J/M^2"
```

```
]RUN
ONE DIMENSIONAL TRANSIENT CONDUCTION
ISOTHERMAL BOUNDARY
HOW MANY LAYERS ARE CHOSEN?
?6
THERE ARE 6 LAYERS
AT WHICH PLANE DO YOU WISH TO
DISPLAY THE TEMPERATURE?
?4
PLANE 4
CHOOSE THE FOURIER NUMBER
?.49
FOURIER NUMBER IS .49
 CONSIDER AN ELAPSED TIME OF 10 MINUTES
TIME STEP IS 68.0555557 SECONDS
NUMBER OF TIME STEPS 9

TEMPERATURE AT PLANE 4 IS 0
TEMPERATURE AT PLANE 4 IS 0
TEMPERATURE AT PLANE 4 IS 0
TEMPERATURE AT PLANE 4 IS 23.5
TEMPERATURE AT PLANE 4 IS 25
TEMPERATURE AT PLANE 4 IS 42
TEMPERATURE AT PLANE 4 IS 43.7
TEMPERATURE AT PLANE 4 IS 56
TEMPERATURE AT PLANE 4 IS 57.7

TEMPERATURE DISTRIBUTION AFTER 10 MINUTES
PLANE 1 T= 100 C
PLANE 2 T= 78.8 C
PLANE 3 T= 66.8 C
PLANE 4 T= 57.7 C
PLANE 5 T= 66.8 C
PLANE 6 T= 78.8 C
PLANE 7 T= 100 C

HEAT TRANSFER IN 10 MINUTES IS 7481666.7 J/M^2


]RUN
ONE DIMENSIONAL TRANSIENT CONDUCTION
ISOTHERMAL BOUNDARY
HOW MANY LAYERS ARE CHOSEN?
?8
THERE ARE 8 LAYERS
AT WHICH PLANE DO YOU WISH TO
DISPLAY THE TEMPERATURE?
?2
PLANE 2
CHOOSE THE FOURIER NUMBER
?.95
FOURIER NUMBER IS .95
 CONSIDER AN ELAPSED TIME OF 10 MINUTES
TIME STEP IS 74.2187501 SECONDS
NUMBER OF TIME STEPS 8

TEMPERATURE AT PLANE 2 IS 0
TEMPERATURE AT PLANE 2 IS 95
TEMPERATURE AT PLANE 2 IS 9.5
TEMPERATURE AT PLANE 2 IS 172.2
TEMPERATURE AT PLANE 2 IS -128.6
TEMPERATURE AT PLANE 2 IS 505.3
TEMPERATURE AT PLANE 2 IS -872.6
TEMPERATURE AT PLANE 2 IS 2322
```

Program nomenclature

N number of layers chosen
E chosen display plane
W width of each layer
A thermal diffusivity
K thermal conductivity
F Fourier number
D timestep
B number of timesteps in 10 minutes
T temperature
X plane number
S timestep number
M temperature sum to calculate heat transfer (program note 7)
Q heat transfer

Program notes

(1) The number of planes is one more than the number of layers. If an even number of planes N is chosen, there will be a centre plane of symmetry for the case of equal wall temperature. This can be seen in the output. The program will accept odd values of N and different temperatures at each boundary (line 270).

(2) The program structure is as follows:

Lines 30–110 choose the number of layers and the display plane.
Line 120 chooses the Fourier number.
Line 150 calculates the size of the timestep.
Line 180 calculates the *approximate* number of time steps in 10 minutes.
Line 200 dimensions the temperature arrays.
Lines 260–300 are the calculation loop for temperatures for each layer.
Lines 324–326 print the final temperature distribution.
Lines 330–390 calculate the heat transfer (J/m^2) during 10 minutes.

(3) Note that by rounding off T(X, S) in line 290 in order to simplify display, calculation errors are induced. If this line is omitted the temperature after approximately 10 minutes at plane 4 is 57.6669 °C and Q = 7 481 549.5 J/m^2. In this case it is a negligible difference but it is a programming fault to round off too soon.

(4) By choosing the number of layers and the Fourier number the time step is set. Since $F \leqslant 0.5$ a large number of layers means a small

time step and more programming time is needed to reach a solution.
(5) The first run with $F = 0.49$ uses the quickest route to solution that is stable (Figure 3.7(b)) but, as shown later in Figure 3.8(a), the temperature updating is uneven. The second run with $F = 0.95$ is unstable and is sketched in Figure 3.7(c).
(6) The final centre line temperature is 57.7 °C; this may be compared with the results in Program 3.3
(7) The calculation of heat transfer is by equation (3.20):

$$Q = \sum_{\text{layer}} mc_p(T_{\text{final}} - T_{\text{zero}})$$

In this case $T_{\text{zero}} = 0$:

$$Q = \sum_{\text{layer}} mc_p T_{\text{final}}$$

and for a layer of thickness w

$$Q = \sum_{\text{layer}} \rho \cdot 1 \cdot wc_p T_{\text{final}} \quad \text{but} \quad \rho c_p = \frac{k}{\alpha}$$

Thus

$$Q = \frac{kw}{\alpha} \sum_{\text{layer}} T_{\text{final}}$$

$$\sum_{\text{layer}} T_{\text{final}} = \left(\frac{T_1 + T_2}{2}\right) + \left(\frac{T_2 + T_3}{2}\right) + \ldots \left(\frac{T_N + T_{N+1}}{2}\right)$$

$$\sum_{\text{layer}} T_{\text{final}} = \left(\frac{T_1 + T_{N+1}}{2}\right) + 2\left(\frac{T_2 + T_3 + \ldots T_N}{2}\right)$$

This is programmed in lines 330–390.
(8) With many layers and a long time interval this method may take a long time to produce suitably accurate results and Program 3.3 may be preferred.
(9) This program can be modified to deal with convective boundaries.
(10) See program notes 4 and 5 of Program 3.3 for a discussion of the relative merits of this explicit method and the implicit method which follows in Program 3.3.

Program 3.3 One-dimensional transient conduction, implicit method

In order to avoid constraints on the value of Fourier number which can cause problems, the implicit forward time difference method may be preferred for the same problem as Program 3.2. The necessary reading for this program is Section 3.7, 3.9 and 3.10.

The Gaussian elimination program is used to solve the $(n + 1)$ equations of the form

$$-F \cdot T_{n-1,1} + (1 + 2F)T_{n,1} - F \cdot T_{n+1,1} = T_{n,0} \quad (3.16)$$

This is done at each time step in turn with the temperatures updated after each step. Although this appears lengthy there is no limitation to the value of F. However care is needed to ensure adequate accuracy, and the two program RUNS show the effect of large timesteps. Program notes 4 and 5 consider these problems in more detail.

```
5  PRINT " ONE DIMENSIONAL TRANSIENT CONDUCTION"
10  PRINT "(IMPLICIT METHOD WITH GAUSSIAN ELIMINATION)
15  REM  CHOOSE A NUMBER OF LAYERS NOT EXCEEDING 20
20  PRINT " PRINT HOW MANY LAYERS ARE CHOSEN?"
25  INPUT N: PRINT "THERE ARE  ";N;" LAYERS"
27  PRINT "AT WHICH PLANE DO YOU WISH TO "
28  PRINT "DISPLAY THE TEMPERATURE?": PRINT : INPUT E: PR
INT "PLANE ";E
30 W = 0.1 / N
35  REM  AA IS THE THERMAL DIFFUSIVITY IN M^2/S
40 AA = 0.000002
45  PRINT "INPUT THE CHOSEN TIME STEP": INPUT D: PRINT "T
IME STEP IS ";D;" SECONDS"
50  PRINT " CONSIDER AN ELAPSED TIME OF 10 MINUTES"
55 B = ( INT ((600 / D) + 0.5))
60  PRINT " NUMBER OF TIME STEPS IS ";B
64 F = (AA * D) / W ^ 2: PRINT "FOURIER NUMBER IS ";F
65  REM  THE WALLS ARE SUDDENLY RAISED IN TEMPERATURE FRO
M 0 C TO 100 C
70  DIM A(21,22): DIM X(21): DIM T(21)
75  REM  SET THE INITIAL TEMPERATURES
80 T(1) = 100:T(N + 1) = 100: FOR C = 2 TO N:T(C) = 0: NE
XT C:G = 1
81  GOTO 100
85  REM  UPDATE THE TEMPERATURES
90  FOR C = 1 TO R:T(C) = X(C): NEXT C:G = G + 1
95  REM  SET MATRIX COEFFICIENTS TO ZERO
100  FOR J = 1 TO N + 1: FOR I = 1 TO N + 2:A(J,I) = 0: N
EXT I: NEXT J
105  REM  SET THE NON ZERO COEFFICIENTS
110 A(1,1) = 1:A(N + 1,N + 1) = 1: FOR C = 2 TO N:A(C,C -
 1) =  - F:A(C,C) = 1 + 2 * F:A(C,C + 1) =  - F: NEXT C
115  REM  SET THE CONSTANTS
120  FOR C = 1 TO N + 1:A(C,N + 2) = T(C): NEXT C
125  REM  SOLVE BY GAUSS,THERE ARE N+1 EQUATIONS
130 R = N + 1
170  FOR J = 1 TO R
180  FOR I = J TO R
190  IF A(I,J) <  > 0 THEN  GOTO 230
200  NEXT I
210  PRINT " SOLUTION NOT POSSIBLE"
220  GOTO 1000
230  FOR K = 1 TO R + 1
240 X = A(J,K)
250 A(J,K) = A(I,K)
```

```
260 A(I,K) = X
270  NEXT K
280 Y = 1 / A(J,J)
290  FOR K = 1 TO R + 1
300 A(J,K) = Y * A(J,K)
310  NEXT K
320  FOR I = 1 TO R
330  IF I = J THEN  GOTO 380
340 Y =  - A(I,J)
350  FOR K = 1 TO R + 1
360 A(I,K) = A(I,K) + Y * A(J,K)
370  NEXT K
380  NEXT I
390  NEXT J
400  FOR P = 2 TO R - 1:X(P) = A(P,R + 1): NEXT P:X(1) =
100:X(R) = 100
410  PRINT "TEMPERATURE AT PLANE ";E;" =";X(E)
415  IF G >  = B THEN  GOTO 1000
420  GOTO 85
1000  END
```

```
]RUN
 ONE DIMENSIONAL TRANSIENT CONDUCTION
(IMPLICIT METHOD WITH GAUSSIAN ELIMINATION)
 PRINT HOW MANY LAYERS ARE CHOSEN?
?6
THERE ARE  6 LAYERS
AT WHICH PLANE DO YOU WISH TO
DISPLAY THE TEMPERATURE?

?4
PLANE 4
INPUT THE CHOSEN TIME STEP
?67
TIME STEP IS 67 SECONDS
 CONSIDER AN ELAPSED TIME OF 10 MINUTES
 NUMBER OF TIME STEPS IS 9
FOURIER NUMBER IS .482399999
TEMPERATURE AT PLANE 4 =3.61350895
TEMPERATURE AT PLANE 4 =9.94290835
TEMPERATURE AT PLANE 4 =17.6014035
TEMPERATURE AT PLANE 4 =25.5930164
TEMPERATURE AT PLANE 4 =33.344347
TEMPERATURE AT PLANE 4 =40.5726916
TEMPERATURE AT PLANE 4 =47.1665404
TEMPERATURE AT PLANE 4 =53.1064813
TEMPERATURE AT PLANE 4 =58.4187987
```

```
]RUN
 ONE DIMENSIONAL TRANSIENT CONDUCTION
(IMPLICIT METHOD WITH GAUSSIAN ELIMINATION)
 PRINT HOW MANY LAYERS ARE CHOSEN?
?6
THERE ARE  6 LAYERS
AT WHICH PLANE DO YOU WISH TO
DISPLAY THE TEMPERATURE?

?4
PLANE 4
INPUT THE CHOSEN TIME STEP
?200
TIME STEP IS 200 SECONDS
 CONSIDER AN ELAPSED TIME OF 10 MINUTES
```

```
 NUMBER OF TIME STEPS IS 3
FOURIER NUMBER IS 1.44
TEMPERATURE AT PLANE 4 =17.4240062
TEMPERATURE AT PLANE 4 =37.2210923
TEMPERATURE AT PLANE 4 =53.7970543

410  PRINT "TEMPERATURE AT PLANE ";E;" =";X(E)
411  PRINT "TEMPERATURE AT PLANE 3= ";X(3)
412  PRINT "TEMPERATURE AT PLANE 2= ";X(2)
413  PRINT


]RUN
 ONE DIMENSIONAL TRANSIENT CONDUCTION
(IMPLICIT METHOD WITH GAUSSIAN ELIMINATION)
 PRINT HOW MANY LAYERS ARE CHOSEN?
?6
THERE ARE  6 LAYERS
AT WHICH PLANE DO YOU WISH TO
DISPLAY THE TEMPERATURE?

?4
PLANE 4
INPUT THE CHOSEN TIME STEP
?67
TIME STEP IS 67 SECONDS
 CONSIDER AN ELAPSED TIME OF 10 MINUTES
 NUMBER OF TIME STEPS IS 9
FOURIER NUMBER IS .482399999
TEMPERATURE AT PLANE 4 =3.61350895
TEMPERATURE AT PLANE 3= 7.35885404
TEMPERATURE AT PLANE 2= 26.3588717

TEMPERATURE AT PLANE 4 =9.94290835
TEMPERATURE AT PLANE 3= 16.5032311
TEMPERATURE AT PLANE 2= 42.0195595

TEMPERATURE AT PLANE 4 =17.6014035
TEMPERATURE AT PLANE 3= 25.5393131
TEMPERATURE AT PLANE 2= 52.2087358

TEMPERATURE AT PLANE 4 =25.5930164
TEMPERATURE AT PLANE 3= 33.8761973
TEMPERATURE AT PLANE 2= 59.4414767

TEMPERATURE AT PLANE 4 =33.344347
TEMPERATURE AT PLANE 3= 41.378479
TEMPERATURE AT PLANE 2= 64.9646045

TEMPERATURE AT PLANE 4 =40.5726916
TEMPERATURE AT PLANE 3= 48.0647569
TEMPERATURE AT PLANE 2= 69.4172655

TEMPERATURE AT PLANE 4 =47.1665404
TEMPERATURE AT PLANE 3= 54.0009609
TEMPERATURE AT PLANE 2= 73.1409452

TEMPERATURE AT PLANE 4 =53.1064813
TEMPERATURE AT PLANE 3= 59.2631365
TEMPERATURE AT PLANE 2= 76.3281159

TEMPERATURE AT PLANE 4 =58.4187987
TEMPERATURE AT PLANE 3= 63.924932
TEMPERATURE AT PLANE 2= 79.0948204
```

Program nomenclature

N number of layers
E display plane chosen for results
W width of each layer
AA thermal diffusivity
D time step
B number of timesteps in 10 minutes (approximately)
F Fourier number
A matrix coefficients for Gaussian elimination
X temperatures calculated by Gaussian elimination
T temperatures before an elimination

Program notes

(1) Rather than choose a Fourier number, here we choose a timestep of any size. To obtain compatibility of results with Program 3.2, six layers are used with a timestep of 67 seconds and the temperature at plane 4 is observed. The resulting temperature is 58.4 °C compared with 57.7 °C by the explicit method after the 10 minute period. A second run with a 200 second timestep yields 53.8 °C, illustrating the need for care (see program note 5).
(2) The program structure is as follows:

Lines 15–64 input of data and determination of the number of timesteps and the Fourier number.
Line 80 setting the initial temperatures.
Lines 85–95 updating temperatures after a timestep.
Line 100 sets all matrix coefficients to zero.
Line 110 sets the non-zero matrix coefficients.
Line 120 sets the matrix constants at the known temperatures.
Lines 125–390 Gaussian elimination routine.
Lines 400–410 print results.
Lines 415–420 if time is elapsed, end the program, if not return to updating temperatures and elimination.

(3) It has not been felt necessary to add the heat transfer calculation of lines 330–390 of Program 3.2. Thermal conductivity data should be given if it is required (2 W/m K).
(4) Figure 3.8(a), left-hand side, shows a plot of the results of Program 3.2. These show irregularity in updating and two almost identical temperatures appear at pairs of timesteps. Indeed if $F = 0.5$ the temperature equation for the explicit method becomes

$$T_{n,1} = \frac{T_{n-1,0} + T_{n+1,0}}{2}$$

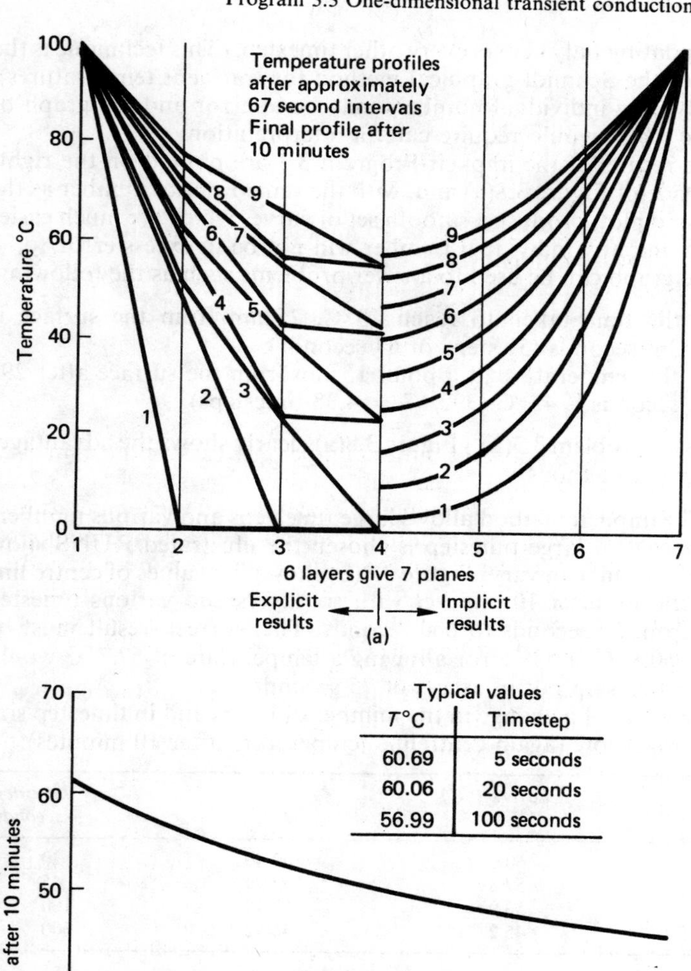
Temperature profiles
after approximately
67 second intervals
Final profile after
10 minutes
Temperature °C
100
80
60
40
20
0
1
2
3
4
5
6
7
8
9
6 layers give 7 planes
Explicit results
Implicit results
(a)
Typical values
T°C
Timestep
60.69
5 seconds
60.06
20 seconds
56.99
100 seconds
Centreline temperature °C after 10 minutes
70
60
50
40
30
0
100
200
300
400
500
600
Timestep (seconds)
(b)

Figure 3.8

and updating only occurs every other timestep. (This technique is the basis of the Schmidt graphical method for transient temperatures). Clearly, any individual numbers may be in error and the graph of Figure 3.8(a) would require careful interpretation.

The results of the implicit Program 3.3 are plotted on the right-hand side of Figure 3.8(a) and, with the same Fourier number as the left-hand plot, produce a smooth set of curves which are much easier to use and any individual number will not be in excess error.

The graph can be used to answer problems such as the following:

(a) the time taken to reach 40 °C, 25 mm from the surface is 369 seconds ($5\frac{1}{2}$ steps of 67 seconds);
(b) the temperature at a point 15 mm from the surface after 293 seconds is 45 °C (392/67 = 4.38 timesteps).

(See also Problem 3.3(a).) Figure 3.8(a) clearly shows the advantages of implicit methods.

(5) The implicit method allows large timesteps and various numbers of layers. If a large timestep is chosen the illustrated RUNS show how the result can vary. Figure 3.8(b) shows the values of centre line temperature after 10 minutes with six layers and various timestep sizes from 20 seconds to 600 seconds. The 'correct' result must be about 60.8 °C, a 5% error allowing a temperature of 57.7 °C would need a timestep not in excess of 75 seconds.

The effect of a change in the number of layers and in timestep size is shown below (again centreline temperature after 10 minutes):

	Layers (°C)		*Timestep*
10	*6*	*4*	*(seconds)*
60.11	60.06	60.40	20
57.86	57.89	57.99	45
56.9	56.99	57.18	100
44.9	45.27	45.92	600

Clearly the number of layers is not as critical as the timestep size.

(6) Since the implicit method tends to overestimate, and the explicit method to underestimate, better results from finite difference temperature calculations would be likely to be achieved by the Crank–Nicholson method (equation (3.17), Section 3.7) using moderate timesteps. This can be achieved by modification of lines 110 and 120 in this program.

Program 3.4 One-dimensional transient conduction with convective boundary

When allowance is made for convection at the surfaces the equations at the wall involve both Fourier number and Biot number. Equation (3.18) is arranged with explicit backward time differences but in view of the discussions in the notes of Program 3.3 this program is written in implicit form with forward time differences when the energy balance at the boundary becomes

$$-kA\frac{(T_{n,1} - T_{n-1,1})}{a} - hA(T_{n,1} - T_f)\Delta t = \rho\frac{a}{2}c_p(T_{n,1} - T_{n,0})$$

which becomes, using $F = \alpha\Delta t/a^2$ and $B = ha/k$,

$$T_{n,1}[1 + 2F + 2FB] - 2FT_{n-1,1} = T_{n,0} + 2FBT_f$$

Thus in the Gaussian elimination method the coefficients of the equations of two surface layers must be changed. The necessary reading for this program is Section 3.8.

The thermal conductivity of the material is 2 W/m K and the surface heat transfer coefficient at each boundary is 120 W/m² K. The *fluid* temperature adjacent to the boundary is 100 °C and the initial material temperature is again 0 °C.

```
5  PRINT " ONE DIMENSIONAL TRANSIENT CONDUCTION"
7  PRINT "WITH CONVECTIVE BOUNDARY"
10  PRINT "(IMPLICIT METHOD WITH GAUSSIAN ELIMINATION)
15  REM  CHOOSE A NUMBER OF LAYERS NOT EXCEEDING 20
20  PRINT " PRINT HOW MANY LAYERS ARE CHOSEN?"
25  INPUT N: PRINT "THERE ARE  ";N;" LAYERS"
27  PRINT "AT WHICH PLANE DO YOU WISH TO "
28  PRINT "DISPLAY THE TEMPERATURE?": PRINT : INPUT E: PR
INT "PLANE ";E
30 W = 0.1 / N
35  REM  AA IS THE THERMAL DIFFUSIVITY,K IS THE THERMAL C
ONDUCTIVITY AND H THE SURFACE HEAT TRANSFER COEFFICIENT
40 AA = 0.000002:H = 120:K = 2
45  PRINT "INPUT THE CHOSEN TIME STEP": INPUT D: PRINT "T
IME STEP IS ";D;" SECONDS"
50  PRINT " CONSIDER AN ELAPSED TIME OF 10 MINUTES"
55 B = ( INT ((600 / D) + 0.5))
60  PRINT " NUMBER OF TIME STEPS IS ";B
64 F = (AA * D) / W ^ 2: PRINT ,"FOURIER NUMBER IS ";F
65  REM  THE WALLS ARE EXPOSED TO A FLUID AT 100 C AT TIM
E ZERO
66 BI = H * W / K: PRINT "BIOT NUMBER IS ";BI
67 TF = 100
70  DIM A(21,22): DIM X(21): DIM T(21)
75  REM  SET THE INITIAL TEMPERATURES
79 G = 1
80  FOR C = 1 TO N + 1:T(C) = 0: NEXT C
81  GOTO 100
85  REM  UPDATE THE TEMPERATURES
90  FOR C = 1 TO R:T(C) = X(C): NEXT C:G = G + 1
95  REM  SET MATRIX COEFFICIENTS TO ZERO
100  FOR J = 1 TO N + 1: FOR I = 1 TO N + 2:A(J,I) = 0: N
EXT I: NEXT J
105  REM  SET THE NON ZERO COEFFICIENTS
110 A(1,1) = 1 + 2 * F + 2 * F * BI:A(N + 1,N + 1) = 1 +
2 * F + 2 * F * BI:A(1,2) =  - 2 * F:A(N + 1,N) =  - 2 *
F: FOR C = 2 TO N:A(C,C - 1) =  - F:A(C,C) = 1 + 2 * F:A(
C,C + 1) =  - F: NEXT C
```

```
115  REM  SET THE CONSTANTS
120  FOR C = 2 TO N:A(C,N + 2) = T(C): NEXT C:A(1,N + 2)
= (2 * F * BI * TF) + T(1):A(N + 1,N + 2) = (2 * F * BI *
 TF) + T(N + 1)
125  REM  SOLVE BY GAUSS,THERE ARE N+1 EQUATIONS
130 R = N + 1
170  FOR J = 1 TO R
180  FOR I = J TO R
190  IF A(I,J) <  > 0 THEN  GOTO 230
200  NEXT I
210  PRINT " SOLUTION NOT POSSIBLE"
220  GOTO 1000
230  FOR K = 1 TO R + 1
240 X = A(J,K)
250 A(J,K) = A(I,K)
260 A(I,K) = X
270  NEXT K
280 Y = 1 / A(J,J)
290  FOR K = 1 TO R + 1
300 A(J,K) = Y * A(J,K)
310  NEXT K
320  FOR I = 1 TO R
330  IF I = J THEN  GOTO 380
340 Y =  - A(I,J)
350  FOR K = 1 TO R + 1
360 A(I,K) = A(I,K) + Y * A(J,K)
370  NEXT K
380  NEXT I
390  NEXT J
400  FOR P = 1 TO R:X(P) = A(P,R + 1): NEXT P
410  PRINT "TEMPERATURE AT PLANE ";E;" =";X(E)
415  IF G >  = B THEN  GOTO 1000
420  GOTO 85
1000  END
```

```
]RUN
 ONE DIMENSIONAL TRANSIENT CONDUCTION
WITH CONVECTIVE BOUNDARY
(IMPLICIT METHOD WITH GAUSSIAN ELIMINATION)
 PRINT HOW MANY LAYERS ARE CHOSEN?
?6
THERE ARE  6 LAYERS
AT WHICH PLANE DO YOU WISH TO
DISPLAY THE TEMPERATURE?

?4
PLANE 4
INPUT THE CHOSEN TIME STEP
?67
TIME STEP IS 67 SECONDS
 CONSIDER AN ELAPSED TIME OF 10 MINUTES
 NUMBER OF TIME STEPS IS 9
FOURIER NUMBER IS .482399999
BIOT NUMBER IS 1
TEMPERATURE AT PLANE 4 =1.30315366
TEMPERATURE AT PLANE 4 =4.14645603
TEMPERATURE AT PLANE 4 =8.17963885
TEMPERATURE AT PLANE 4 =12.9355087
TEMPERATURE AT PLANE 4 =18.0255268
TEMPERATURE AT PLANE 4 =23.179907
TEMPERATURE AT PLANE 4 =28.23022
TEMPERATURE AT PLANE 4 =33.0801782
TEMPERATURE AT PLANE 4 =37.6803864
```

Program nomenclature

As Program 3.3 with, additionally,
K thermal conductivity of material in slab
H surface heat transfer coefficient
TF fluid temperature

Program notes

(1) The amendments to Program 3.3 mainly consist of the following:

Line 7 new title.
Lines 35 and 40 values of thermal conductivity and surface heat transfer coefficient.
Line 65 REM statement change.
Lines 66 and 67 fluid temperature set, Biot number determined.
Line 110 coefficients of A(1, 1), A(N + 1, N + 1) and A(1, 2), A(N + 1, N) set for convective boundaries.
Line 120 constants A(1, N + 2) and A(N + 1, N + 2) set for convective boundaries.

(2) The problem can be solved by Schmidt graphical plot or transient heat transfer chart giving centre line temperatures of approximately 33 °C and 37 °C respectively.
(3) The program is completely stable and time interval and layer thickness can be chosen freely, but as in Program 3.3 care is needed in selection to achieve the desired accuracy.

Program 3.5 Two-dimensional transient conduction, explicit method

This program is a two-dimensional version of Program 3.2 and suffers the same drawbacks discussed in Program 3.3, notes 4 and 5. A rectangular block has equal length sides which are all raised at time zero from 0 °C to 100 °C. Since heating is on four sides the centre temperature will become higher in 10 minutes than the slab centre line temperature. The stability criterion becomes more stringent with $F \leqslant 0.25$ if the equivalent to equation (3.15) is developed. In order to check the results the transient temperature charts used for slab solutions can be used by the technique of product solution (Figure 3.9) [1]. This shows that

$$\theta_{x,y,t} = (\theta_{x,t} \times \theta_{y,t})$$

where

$$\theta_{x,y,t} = \frac{T_{x,y,t} - T_{\text{fluid}}}{T_{\text{initial}} - T_{\text{fluid}}}$$

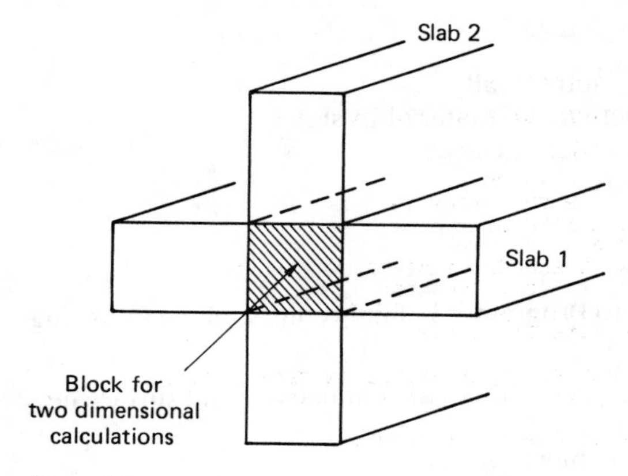

Figure 3.9

and

$$\theta_{x,t} = \frac{T_{x,t} - T_{\text{fluid}}}{T_{\text{initial}} - T_{\text{fluid}}} \quad \text{and} \quad \theta_{y,t} = \frac{T_{y,t} - T_{\text{fluid}}}{T_{\text{initial}} - T_{\text{fluid}}}$$

Isothermal wall temperatures occur if $1/B = k/ha = 0$. In this case the wall temperature is isothermal and we have the solution 57.5 °C for such a problem (Program 3.2):

$$\theta_{x,t} = \theta_{y,t} = \frac{57.5 - 100}{0 - 100} = 0.423$$

which can be checked on a suitable chart. Hence

$$\theta_{x,y,t} = (0.423)^2 = 0.179 = \frac{T_{x,y,t} - 100}{0 - 100}$$

$$T_{x,y,t} = 82.1\,°\text{C}$$

The necessary reading for this program is Sections 3.7, 3.9 and 3.11.

Reference

1. Ozisik, M.N. *Heat Transfer*, McGraw-Hill (1985).

```
10   PRINT "TWO DIMENSIONAL TRANSIENT CONDUCTION"
20   PRINT "ISOTHERMAL BOUNDARY"
30   REM  CHOOSE AN EVEN NUMBER OF LAYERS NOT EXCEEDING 10
.THERE ARE THE SAME NUMBER IN EACH DIMENSION AND 10 LAYER
S HAVE 11 PLANES.
40   REM  AN EVEN NUMBER ENSURES A CENTRE POINT
50   PRINT " HOW MANY LAYERS ARE CHOSEN?"
60   INPUT N: PRINT "THERE ARE ";N;" LAYERS"
```

```
70  REM  THIS PROGRAM WILL DISPLAY THE CENTRE POINT TEMPE
RATURE
80  REM  W IS THE STRIP WIDTH
90 W = 0.1 / N
100  REM  A IS THE THERMAL DIFFUSIVITY IN M^2/S
110 A = 0.000002
120  PRINT "CHOOSE THE FOURIER NUMBER"
125  REM  FOR STABILITY F<=0.25
130  INPUT F: PRINT "FOURIER NUMBER IS ";F
140  REM  DETERMINE THE TIME STEP
150 D = F * (W ^ 2) / A
160  PRINT " CONSIDER AN ELAPSED TIME OF 10 MINUTES"
170  PRINT "TIME STEP IS ";D;" SECONDS"
180 B = 600 / D:B = ( INT (B + 0.5))
190  PRINT "NUMBER OF TIME STEPS ";B
200  DIM T(11,11,B)
210  REM   THE WALLS ARE SUDDENLY RAISED IN TEMPERATURE F
ROM 0 C TO 100 C
220  PRINT
230  FOR Y = 1 TO N + 1: FOR X = 1 TO N + 1
240 T(Y,X,0) = 0
250  NEXT X: NEXT Y
260  FOR S = 1 TO B: FOR X = 1 TO N + 1
270 T(1,X,S) = 100:T(N + 1,X,S) = 100
280  NEXT X
290  FOR Y = 1 TO N + 1
300 T(Y,1,S) = 100:T(Y,N + 1,S) = 100
310  NEXT Y: NEXT S
320  FOR S = 1 TO B: FOR Y = 2 TO N: FOR X = 2 TO N
330 T(Y,X,S) = F * (T(Y + 1,X,S - 1) + T(Y - 1,X,S - 1) +
 T(Y,X + 1,S - 1) + T(Y,X - 1,S - 1) + ((1 / F) - 4) * T(
Y,X,S - 1))
340 T(Y,X,S) = ( INT (T(Y,X,S) * 10 + 0.5)) / 10
350  NEXT X: NEXT Y
370  PRINT "TEMPERATURE AT CENTRE POINT IS ";T((N / 2) +
1,(N / 2) + 1,S);" C"
380  NEXT S
400  REM   A SYMMETRY TEST IS USED AS A CHECK BETWEEN COL
UMN 2 AND ROW 2 AT TIME 5
410  PRINT
420  PRINT "SYMMETRY TEST"
430  PRINT
450  FOR Z = 1 TO N + 1
460  PRINT  TAB( 4)T(2,Z,5), TAB( 9)T(Z,2,5)
470  NEXT Z

]RUN
TWO DIMENSIONAL TRANSIENT CONDUCTION
ISOTHERMAL BOUNDARY
 HOW MANY LAYERS ARE CHOSEN?
?6
THERE ARE 6 LAYERS
CHOOSE THE FOURIER NUMBER
?.24
FOURIER NUMBER IS .24
 CONSIDER AN ELAPSED TIME OF 10 MINUTES
TIME STEP IS 33.3333334 SECONDS
NUMBER OF TIME STEPS 18

TEMPERATURE AT CENTRE POINT IS 0 C
TEMPERATURE AT CENTRE POINT IS 0 C
TEMPERATURE AT CENTRE POINT IS 0 C
TEMPERATURE AT CENTRE POINT IS 5.6 C
TEMPERATURE AT CENTRE POINT IS 14.1 C
TEMPERATURE AT CENTRE POINT IS 24.1 C
TEMPERATURE AT CENTRE POINT IS 32.7 C
TEMPERATURE AT CENTRE POINT IS 41.4 C
```

```
TEMPERATURE AT CENTRE POINT IS 48.5 C
TEMPERATURE AT CENTRE POINT IS 55.3 C
TEMPERATURE AT CENTRE POINT IS 60.9 C
TEMPERATURE AT CENTRE POINT IS 66 C
TEMPERATURE AT CENTRE POINT IS 70.2 C
TEMPERATURE AT CENTRE POINT IS 74.1 C
TEMPERATURE AT CENTRE POINT IS 77.4 C
TEMPERATURE AT CENTRE POINT IS 80.4 C
TEMPERATURE AT CENTRE POINT IS 82.8 C
TEMPERATURE AT CENTRE POINT IS 85.1 C

SYMMETRY TEST

   100          100
   75.7         75.7
   60           60
   54.3         54.3
   60           60
   75.7         75.7
   100          100
```

Program nomenclature

N number of layers
W strip width
A thermal diffusivity
F Fourier number
D timestep
B number of timesteps in 10 minutes (approximately)
T temperature
X, Y position of point in the plane
S timestep number
Z values of X and Y in symmetry test

Program notes

(1) An *even* number of layers is chosen to achieve a centre point.
(2) The program structure is as follows:

Lines 30–110 choose the number of layers, determine layer size.
Line 120 chooses the Fourier number.
Lines 140–190 determine the timestep and number of steps.
Lines 200–380 calculate all temperatures and display the centre temperature at each timestep.
Lines 400–700 are a symmetry test displaying temperatures across the second row beside those down the second 'column'.

(3) Again rounding off errors may be induced by line 340.
(4) A similar program could be used for three-dimensional problems but could cause storage problems. It would be possible not to store all the data if it was not required as only the previous timestep data is used in the calculations.

(5) By similar arguments to those for choosing implicit methods for one-dimensional problems, implicit programs may be preferred for two- or three-dimensional solutions.

PROBLEMS

(3.1)

(a) Rearrange Problem 3.1 for Gauss–Seidel iteration with the temperatures updated as they are found.
(b) A more efficient way of entering data in a matrix with many zero values is to set all the coefficients to zero and then amend the non-zero values. Write a program to achieve this for the Gaussian elimination method of Program 3.1.
(c) Rearrange Program 3.1 to allow for convective boundaries using equations (3.9), (3.11) and (3.12). Evaluate the program with an inside flue gas temperature of 300 °C, an outside air temperature of 10 °C with surface heat transfer coefficients of 18 W/m^2 K on each side.
(d) A rectangular conductor is 0.04 m by 0.03 m and carries an electric current of density 20 000 amp/m^2. The thermal conductivity of the conductor is 70 W/m K and the electrical resistivity is 4.1×10^{-7} ohm m. The conductor is cooled by forced air convection giving a surface heat transfer coefficient of 25 W/m^2 K. The air temperature is 10 °C. Write a program to determine the temperature distribution in the conductor.

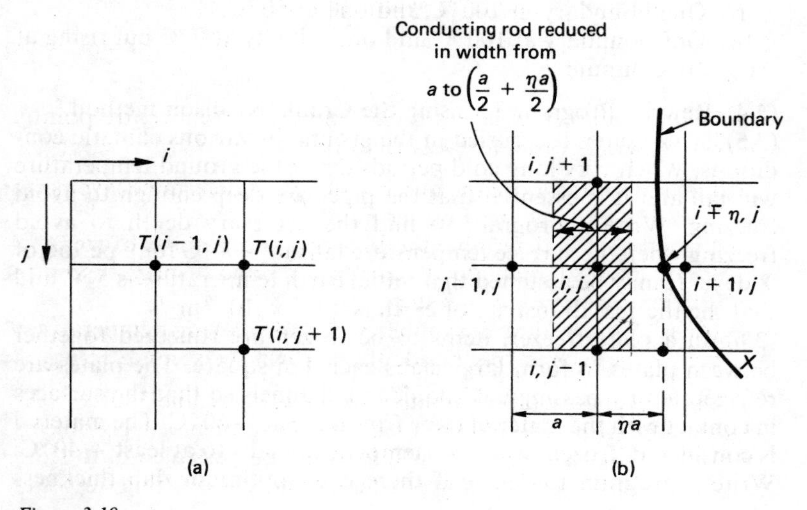

Figure 3.10

(3.2) A more general approach to two-dimensional finite difference problems would be to designate each point (i, j), as in Figure 3.10(a), and rewrite equations in the form

$$T(i-1,j) + T(i+1,j) + T(i,j-1) + T(i,j+1) - 4T(i,j) = 0$$

This type of programming can be used for both steady state and transient problems. Void spaces and curved boundaries would need special consideration. Curved boundaries are considered in Problem 3.8. Reconsider Problem 3.1(c) by this method and then rewrite Program 3.5 using the same technique.

(3.3)

(a) Programs 3.2, 3.3 and 3.4 will give specific answers to two types of problem:
(i) Determine the time taken to reach a temperature T_1 at a point x_1 mm from the boundary.
(ii) Determine the temperature at a point x_2 from the boundary after t seconds. Rearrange the programs to supply answers to these problems with $T_1 = 27\,°C$, $x_1 = 23$ mm, $x_2 = 46$ mm and $t = 248$ seconds.

(b) Rearrange Problem 3.3 to allow for different boundary conditions:
(i) One boundary at 100 °C and one at 50 °C.
(ii) One boundary at 100 °C and one initially at 0 °C but rising at 10 K/minute.

(3.4) Rewrite Program 3.3 using the Crank–Nicolson method.
(3.5) Water pipes are buried in the ground in various climatic conditions. When very long cold periods occur the ground temperature will fall and it is essential that the pipes are deep enough to avoid freezing. Write a program to find the necessary depth to avoid freezing when the surface temperature falls to $-X$°C for a period of Y days. It may be assumed that initial earth temperature is 5 °C and that the thermal diffusivity of earth is $0.14 \times 10^{-6}\,m^2/s$.
(3.6) In a plate freezer, items to be frozen are squeezed together between plates to form large slabs each 1 m square. The plates are refrigerated by passing cold liquids in channels so that the surfaces in contact with the material to be frozen are at −60 °C. The material is considered 'frozen' when the temperature falls to at least −40 °C. Write a program to decide if there is an optimum slab thickness

which will enable material to be processed at the maximum "mass flow rate". If there is such an optimum (or if not use a thickness of 0.3 m) consider the convection problems (Chapter 4) of removing the energy from the freezer plates to maintain them at $-60\,°C$ using a fluid available at $-80\,°C$:

Material properties: $\rho = 220\,kg/m^3$, $k = 0.6\,W/m\,K$ and $c_p = 3\,kJ/kg\,K$
Refrigerant fluid properties in the temperature range -60 to $-80\,°C$: $\rho = 1150\,kg/m^3$, $c_p = 4.3\,kJ/kg\,K$, $k = 0.52\,W/m\,K$ and $\mu = 2.1\,g/s\,m$.

(3.7) A long, thick steel plate leaves a furnace at V m/s at a uniform temperature of 500 °C. The steel is cooled by forced convection and radiation to the air surroundings at 30 °C and the *combined* surface heat transfer coefficient is $450\,W/m^2\,K$. The plate is 30 mm thick and has a thermal conductivity of 49 W/m K, a thermal diffusivity of $1 \times 10^{-5}\,m^2/s$ and a density of $7360\,kg/m^3$. Write a program to determine the surface temperature and the centre line temperature at 1 m intervals from the furnace exit plane for various exit speeds from 0.1 to 1 m/s. Conduction across and along the plate may be neglected.
(3.8) Many problems present curved boundaries for which new finite difference equations must be developed. Consider the case shown in Figure 3.10(b) in which the node $(i + 1, j)$ is *outside* the solid.

In the steady state with $\dot{Q}''' = 0$:

$$ka\left(\frac{1}{2} + \frac{\eta}{2}\right)\left(\frac{T(i, j+1) - T(i,j)}{a}\right) + ka\left(\frac{T(i-1, j) - T(i,j)}{a}\right)$$
$$+\, ka\left(\frac{1}{2} + \frac{\eta}{2}\right)\left(\frac{T(i, j-1) - T(i,j)}{a}\right)$$
$$+ka\left(\frac{T(i+\eta, j) - T(i,j)}{\eta^a}\right) = 0$$

Hence

$$\eta\left(\frac{1+\eta}{2}\right)(T(i, j+1) + T(i, j-1)) + \eta T(i-1, j)$$
$$+\, T(i+\eta, j) - (1+\eta)^2\, T(i,j) = 0$$

The point X in Figure 3.10(b) will involve a double effect since both the i and j grid lines will be shortened above and to the right of X. Develop an equation for this type of point.

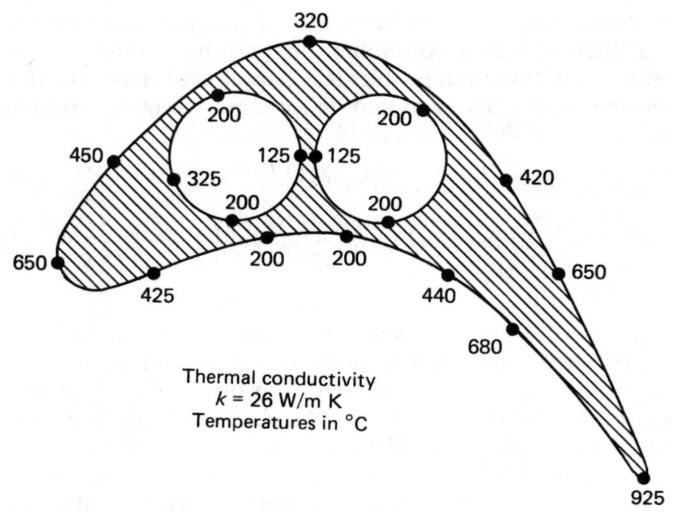

Figure 3.11

With these two new equations and those developed before write a program to determine the steady state temperature distribution in the cooled gas turbine blade shown in Figure 3.11, which is to scale.

Open-ended problems

(3.9) A proposed electric floor heating system for a room 6 m × 5 m has electric wires arranged uniformly under a layer of concrete so that 5 kW of energy is transferred uniformly to the *lower* concrete surface. The heating system is controlled and is set initially to be turned on or off every 6 hours so that there is a total 'on' time of 12 hours in a 24 hour period. The room is to be maintained at 19 °C when the heat losses through the fabric vary between 0 and 2.5 kW depending on the outside temperature.

Cheaper 'off-peak' electricity is available between 11 p.m. and 7 a.m. and between 1 p.m. and 4 p.m. To minimise costs it is preferable that advantage should be taken of this facility, provided that the desired room state can be achieved.

The surface heat transfer coefficient allowing for both convection and radiation from the floor is 10 W/m^2 K, the thermal diffusivity of concrete is 8×10^{-7} m^2 /s and the thermal conductivity of concrete is 0.8 W/m K. Write a program to find out how to set the heating system controller to achieve the best approach to the target for various rates of heat loss through the fabric with different floor concrete thicknesses. It may well be that some additional control features may be required to achieve the objective.

(3.10) There is no reason why the techniques discussed in this chapter should not be extended to three-dimensional problems (except for the computer memory size). Pose a 'real' problem of interest involving this advance in application (cylinder block in a car engine, a gas turbine blade, etc.). If you have access to a suitable piece of software which will solve such a problem you should be able to check the validity of your own program.

Chapter 4

Convection

4.1 Introduction

Simple concepts of convection were introduced in Chapter 2 where it was stated that the determination of the surface heat transfer coefficient h for use in the relation

$$\dot{Q}'' = h\theta$$

was complex because the values depended on the fluid properties, the flow pattern and the geometry of the bounding surface. This chapter is concerned with methods of determining h.

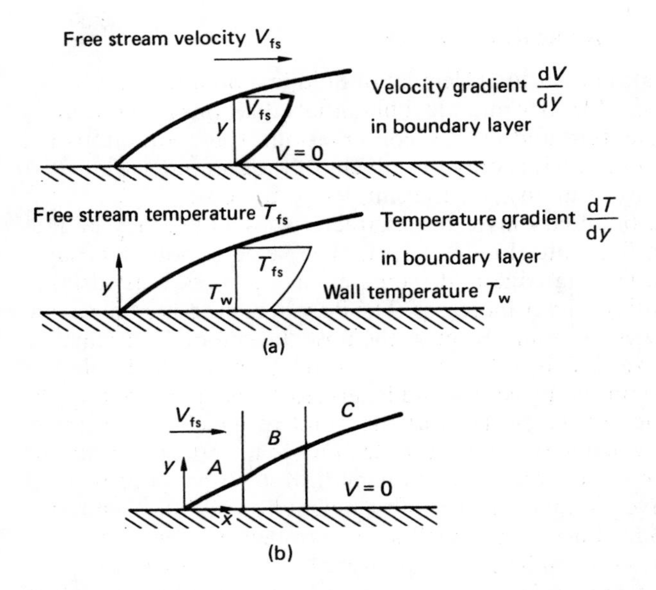

Figure 4.1

4.2 Boundary layers

When a fluid flows over a solid boundary the fluid particles immediately adjacent to the boundary are at rest and there is a velocity gradient in the thin film of fluid close to the boundary as the velocity changes from zero to the free stream value. This thin film is called the hydrodynamic boundary layer. A similar situation occurs in the thin film known as the *thermal boundary layer* in which the fluid temperature changes from the wall temperature to the free stream temperature. The thickness of these two boundary layers is not equal. Figure 4.1(a) shows the gradients in these boundary layers. In order to determine h an analysis of the mass, momentum and energy balances in these layers is required and the importance of this need is emphasised by the wide range of surface heat transfer coefficients which are found.

Free convection	gases 0.5–500 W/m^2 K liquids 50–2000 W/m^2 K
Forced convection	gases 10–700 W/m^2 K liquids 100–10 000 W/m^2 K
Phase change	100–100 000 W/m^2 K

4.3 Boundary layer analysis

By consideration of a small element of fluid in a boundary layer (δx, δy, δz) it is possible to formulate differential equations for continuity (mass), momentum and energy conservation. These equations are difficult to solve and it is only the advent of the computer which has enabled solutions in any but the simplest situations.

Flow in a boundary layer has characteristics which may be illustrated for a flat plate (Figure 4.1(b)). The velocity boundary layer starts at the leading edge and increases in thickness y as distance from the leading edge x increases. The flow in zone A is smooth and termed *laminar*. At some point in the flow the smoothness starts to change and zone B is a transition to zone C in which the flow is randomly disturbed by eddies and is termed *turbulent* (note that the thermal boundary layer starts at the point at which heat transfer commences which may not be at the leading edge of the plate, adding additional complication). The mathematical description of laminar flow is relatively simple but that of turbulent flow is difficult because of the random eddies. The modelling of turbulence is very important to any solution, including those performed by computer.

Because of the problems in differential analysis solutions alternative approaches have been used. By considering a control volume across the whole boundary layer, integral equations for the conservation of mass, momentum and energy can be formulated and, with *empirical* knowledge of boundary layer temperature and velocity profiles represented by quartic or, more commonly, cubic, equations, solved for laminar, forced flow over simple surfaces.

Neither the differential nor the integral approaches will be discussed further here but both are well documented [1].

4.4 Determination of laminar and turbulent flow

Experiment shows that the transition from laminar to turbulent flow occurs at predictable conditions. In forced flow the Reynolds number, Re, is the parameter which indicates the flow pattern

$$\mathrm{Re} = \frac{\rho V l}{\mu}$$

where ρ is fluid density
V is fluid free stream velocity
μ is fluid dynamic viscosity
and l is a representative length dimension: diameter of a tube, distance from the leading edge for a plate, and for non-circular ducts the hydraulic diameter D_h where

$$D_h = \frac{4 \times \text{area of cross-section}}{\text{wetted perimeter}}$$

For a plane surface in forced flow:

$$\mathrm{Re} < 500\,000 \text{ flow is laminar}$$

$$\mathrm{Re} > 500\,000 \text{ flow is turbulent}$$

For a tubular surface in forced flow:

$$\mathrm{Re} < 2000 \text{ flow is laminar}$$

$$\mathrm{Re} > 2300 \text{ flow is becoming turbulent}$$

but there is divergence of opinion for the value of Re when turbulence is fully established. Special situations occur in transition zones between laminar and turbulent flow and in entry lengths of tubes where the boundary layer develops from the whole of the surrounding walls and eventually meets in the middle to produce fully developed flow. Empirical data is available for these situations (Figure 4.2) which will state the ranges of Re to which the data is applicable [2].

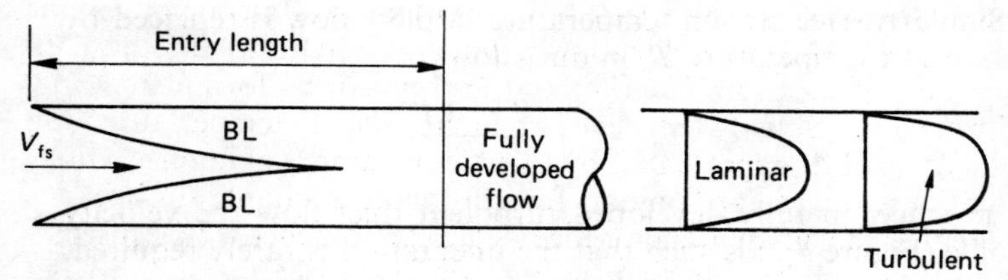

Figure 4.2

4.5 Reynolds' analogy

Because of the historical difficulty with turbulence, Osborne Reynolds, who observed the similarity between momentum transfer and heat transfer, postulated a simple analogy which enabled heat transfer coefficients to be predicted from a measurement of pressure loss due to friction in the flow. The analogy has been considerably amended to account for increasing knowledge of boundary layers. The development of the Reynolds analogy relation leads to the result

$$\frac{\mathrm{Nu}}{\mathrm{Re, Pr}} = \mathrm{St} = \frac{f}{2} \tag{4.1}$$

where Nu is the Nusselt number $\frac{hl}{k}$

Pr is the Prandtl number $\frac{\mu c_p}{k}$

and St is the Stanton number $\frac{h}{\rho V c_p}$

The quantity f is the friction factor defined by

$$f = \frac{\tau_{wall}}{\frac{1}{2}\rho V^2} \tag{4.2}$$

available by experiment, chart or by calculation from relations such as the Blasius equation for smooth tubes:

$$f = 0.079\ \mathrm{Re}^{-0.25} \tag{4.3}$$

τ_{wall} is the shear stress at the wall and in all these relations V is interpreted as the free stream velocity for plate flow and the bulk velocity V_b in duct flow where

$$V_b = \int_A \frac{V \cdot dA}{A}$$

Similarly, free stream temperature in plate flow is replaced by bulk mean temperature T_b in duct flow:

$$T_b = \int_A \frac{TV \cdot dA}{V_b}$$

In practice, in fully developed turbulent duct flow the velocity profile (Figure 4.2) is such that the integration is rarely required.

Reynolds analogy may be applied to forced turbulent flow provided Pr = 1. In most situations this is not the case and several modifications are available. The simplest is the Colburn modification valid for $0.6 < \text{Pr} < 50$:

$$\text{St} \cdot \text{Pr}^{2/3} = \frac{f}{2} \tag{4.4}$$

another is the Prandtl–Taylor modification for all Pr:

$$\text{St} = \frac{f}{2}\left[\frac{1}{1 + r_v(\text{Pr} - 1)}\right] \tag{4.5}$$

This latter relation (equation (4.5)) is based on a two-part boundary layer including a laminar sublayer below the turbulent layer (Figure 4.3). The quantity r_v is the ratio between the velocity at the edge of the sublayer and the bulk or free stream velocity and is found by experiment:

$$r_v = \frac{V_l}{V_{fs}} = 2.11\ \text{Re}^{-0.1} \text{ for smooth plates}$$

$$r_v = \frac{V_l}{V_b} = 1.99\ \text{Re}^{-1/8} \text{ for smooth tubes}$$

It should be clear that in all cases h is found by interpreting St. Since

$$\text{St} = \phi\left(\frac{f}{z}\right)$$

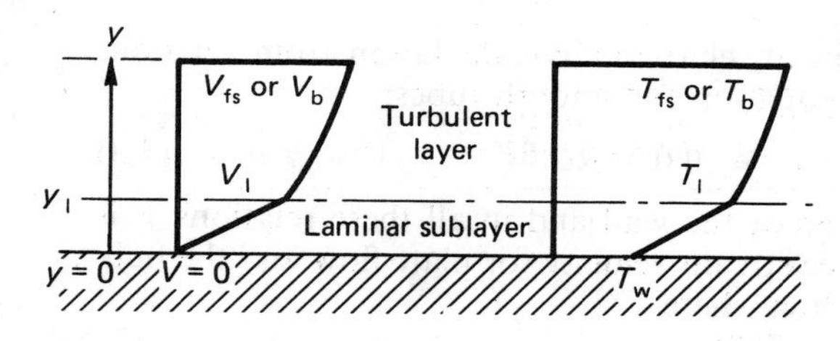

Figure 4.3

$$h = \rho V c_p \phi\left(\frac{f}{2}\right) \qquad (4.6)$$

4.6 Dimensional analysis in convection

In view of the problems of analytical approaches many of the equations to determine h are based on dimensional analysis backed by experimental work. In *forced* convection the dimensional analysis shows that

$$\text{Nu} = \phi(\text{Re}, \text{Pr}) \qquad (4.7)$$

and the experimental results give the actual formulation. When determining the values of Nu, Re and Pr, fluid properties are involved in a boundary layer in which temperature varies. The fluid properties μ, k, ρ and c_p fall into two groups. Bulk properties ρ and c_p, which are concerned with the enthalpy of the whole flow of fluid, are usually evaluated at the bulk or free stream temperature, whereas transport properties, μ and k, which are concerned with momentum and energy transfer in regions of velocity and temperature gradient, are sometimes evaluated at the film temperature T_f. Film temperature is the arithmetic mean between the bulk or free stream temperature, T_b and the wall temperature T_w:

$$T_f = \frac{T_b + T_w}{2}$$

When using any empirical equation care must be taken to use the prescribed temperatures to evaluate the fluid properties. Values of fluid properties needed for convective heat transfer for air and water are included in the programs in this chapter.

A few typical relations are tabulated below for plane surfaces and tubes in laminar and turbulent flow. These give *average* Nu for finite lengths of surface. If other relations are required specialist texts should be consulted [2].

Surface	*Laminar flow*	*Turbulent flow*	*Properties at*	
Plane	$\text{Nu} = 0.664\,\text{Re}^{0.5}\,\text{Pr}^{0.33}$	$\text{Nu} = 0.036\,\text{Re}^{0.8}\,\text{Pr}^{0.33}$	Film temperature	(4.8)
Tube	$\text{Nu} = 3.66$ (long tube)	$\text{Nu} = 0.023\,\text{Re}^{0.8}\,\text{Pr}^{0.4}$	Bulk temperature	

4.7 Free convection

The application of dimensional analysis to free convection shows that

$$\mathrm{Nu} = \phi(\mathrm{Pr}, \mathrm{Gr})$$

where GR is the Grashof number

$$\mathrm{Gr} = \frac{\rho^2 \beta g \theta l^3}{\mu^2}$$

β is the coefficient of cubical expansion ($\beta = 1/T$ for a perfect gas)
g g is gravitational acceleration
T is thermodynamic temperature
and θ is a characteristic temperature difference

The product of Pr and Gr is called the Rayleigh number (Ra) and it is found that in a similar way that Re dictates the onset of turbulence in forced flow, so Ra is the parameter in free convection. For *plane or cylindrical* vertical surfaces it is found that:

Flow	*Rayleigh number*	*Relation*	
Laminar	$\mathrm{Ra} < 10^9$	$\mathrm{Nu} = 0.59(\mathrm{Pr}, \mathrm{Gr})^{1/4}$	(4.9)
Turbulent	$\mathrm{Ra} > 10^9$	$\mathrm{Nu} = 0.13(\mathrm{Pr}, \mathrm{Gr})^{1/3}$	

The representative length is height and the Nu are average values over the surface. Film temperature is used for properties and θ is the difference between surface temperature and fluid free stream temperature. Results for other situations are found in the references [2].

4.8 Phase change

The table in Section 4.2 showed that phase change heat transfer coefficients can be very high. Laminar *film* consideration can be analysed by considering the equilibrium of the flow of condensate under the action of gravity and viscosity, and relations for inclined vertical surfaces and the exterior of horizontal tubes can be deduced (Program 4.5). Evaporation in tubes and *drop wise* condensation are not considered but Problems 4.11 and 4.13 consider other phase change situations. Information can be found in the references [1, 2, 3].

4.9 Complex surfaces

Complex surfaces used in heat exchangers such as car radiators do not lend themselves to analysis. However the heat transfer coefficients will be governed by equation (4.7) and the results of experiments can be presented graphically (Figure 7.5).

4.10 Use of empirical relations

The relations presented here are often based on experimental work and can lead to considerable error in the prediction of heat transfer coefficients. For example, the relation in equation (4.8):

$$\mathrm{Nu} = 0.023\,\mathrm{Re}^{0.8}\,\mathrm{Pr}^{0.4}$$

is reported to give errors of up to 25%. Other relations are available for similar situations based on more modern investigations. For example, a relation by Petukhov has errors of between 5 and 10% (Problem 4.8) [4]. Considerable care must be taken in allowance for such variations in any design calculation.

References

1. Eckert, E.R.G. and Drake, R.M. *Analysis of Heat and Mass Transfer*, McGraw-Hill (1972).
2. Ozisik, M.N. *Heat Transfer*, McGraw-Hill (1985).
3. Kreith, F. and Bohn, M.S. *Principles of Heat Transfer*, Harper and Row (1986).
4. Petukhov, B.S. 'Heat transfer and friction in turbulent pipe flow with variable physical properties'. In *Advances in Heat Transfer*, Vol. 6, pp. 504–64, Academic Press (1970).

WORKED EXAMPLES

Program 4.1 Water and air properties

Fluid properties are required to determine surface heat transfer coefficients. This program presents a data block for water which is interpreted using the Lagrangian routine of Program 1.1. The program is arranged for 11 sets of water data from 0 °C to 100 °C but a second block of 10 sets of data for air at atmospheric pressure from −100 °C to 800 °C is appended [1].

The data in each line consists of six numbers:

(a) temperature, T, in °C;
(b) specific volume, v, in m³/kg ($v = 1/\rho$);
(c) specific heat capacity at constant pressure, c_p in kJ/kg K;
(d) thermal conductivity, k, in W/m K;
(e) dynamic viscosity, μ, in g/m s;
(f) Prandtl number, Pr.

Reference

1. Haywood, R.W. *Thermodynamic Tables in S.I. (Metric) Units*, Cambridge University Press (1968).

```
10  PRINT "TITLE"
20  REM  THE PROGRAM USES THE LAGRANGE INTERPOLATION FOR
WATER PROPERTIES
30  DIM T(11),VS(11),CP(11),K(11),MU(11),PR(11)
40  FOR N = 0 TO 10
50  READ T(N),VS(N),CP(N),K(N),MU(N),PR(N)
60  NEXT N
65  REM  THIS PROGRAM NEEDS AN INPUT "A" TO SET THE TEMPE
RATURE AT WHICH THE WATER PROPERTIES ARE REQUIRED...SEE L
INE1050
100 A = 60
1000 B = 0:C = 0:E = 0:F = 0:G = 0
1010  FOR J = 0 TO 10
1020 X = 1
1030  FOR N = 0 TO 10
1040  IF N = J THEN  GOTO 1060
1050 X = X * (A - T(N)) / (T(J) - T(N))
1060  NEXT N
1070 B = B + X * VS(J)
1080 C = C + X * CP(J)
1090 E = E + X * K(J)
1100 F = F + X * MU(J)
1110 G = G + X * PR(J)
1115  NEXT J
1120 VS = B:CP = C:K = E:MU = F / 1000:PR = G
2000  DATA  0,0.001,4.217,0.569,1.755,13.02
2010  DATA  10,0.001,4.193,0.587,1.301,9.29
2020  DATA  20,0.001,4.182,0.603,1.002,6.95
2030  DATA  30,0.001,4.179,0.618,0.797,5.39
2040  DATA  40,0.00101,4.179,0.632,0.651,4.31
2050  DATA  50,0.00101,4.181,0.643,0.544,3.53
2060  DATA  60,0.00102,4.185,0.653,0.462,2.96
2070  DATA  70,0.00102,4.19,0.662,0.4,2.53
2080  DATA  80,0.00103,4.197,0.67,0.35,2.19
2090  DATA  90,0.00104,4.205,0.676,0.311,1.93
2100  DATA  100,0.00104,4.216,0.681,0.278,1.273
2200  REM  THIS IS A DATA BLOCK FOR AIR PROPERTIES WHICH
MAY BE ADDED TO THE WATER PROPERTIES PROGRAM WITH THE WAR
NING THAT THERE ARE ONLY 10 ROWS INSTEAD OF 11 AND THAT S
OME CHOICE LINES WOULD NEED TO BE ADDED TO SELECT WATER O
R AIR
2210  DATA  -100,0.488,1.01,0.016,0.012,0.75
2220  DATA  0,0.773,1.01,0.024,0.017,0.72
2230  DATA  100,1.057,1.02,0.032,0.022,0.7
2240  DATA  200,1.341,1.03,0.039,0.026,0.69
2250  DATA  300,1.624,1.05,0.045,0.03,0.69
2260  DATA 400,1.908,1.07,0.051,0.033,0.7
2270  DATA  500,2.191,1.1,0.056,0.036,0.7
2280  DATA  600,2.473,1.12,0.061,0.039,0.71
2290  DATA  700,2.756,1.14,0.066,0.042,0.72
2300  DATA  800,3.039,1.16,0.071,0.044,0.73
```

Program nomenclature

T	temperature
VS, B	specific volume
CP, C	specific heat capacity at constant pressure
K, E	thermal conductivity
MU, F	dynamic viscosity
PR, G	Prandtl number
A	temperature at which property is required

Program notes

(1) The program produces values of data at the temperature A in line 1120. It would be wise to test the output by temporarily adding lines 1130 onwards asking for a print out of the data at temperature A which can be checked against tables before incorporating this routine into a program.
(2) Make a list of all the variables used before incorporating this routine into a program so that they are not used in the program to represent other items.
(3) Any program in which this routine is directly incorporated would have to fit in the spaces between lines 100 and 1000 or between 1120 and 2300. If in doubt about length start the program numbering *after* the last data block.
(4) The program has not been incorporated into all the relevant programs to save space. It is used in some and Program 4.3 may be examined as an example.

Program 4.2 Experimental Nu–Re relation

Figure 4.4 shows a tube through which air is blown. The air is heated by an electric element wound uniformly around the outside of the tube and is well insulated so that all the heater input is assumed to pass to the air flow. The tube has a thermocouple fitted to measure the wall temperature at the centre point 0.91 m from the ends. The air flow is measured by an orifice pressure drop and the pressure and temperature at the orifice are measured. Five sets of readings are given at different air flow rates. Write a program to determine the surface heat transfer coefficient, the Nusselt and Reynolds numbers and, using the LINEFIT Program 1.3, determine a relation for Nu = ϕ(Re, Pr) assuming that Nu $\propto$ $\mathrm{Pr}^{0.4}$.

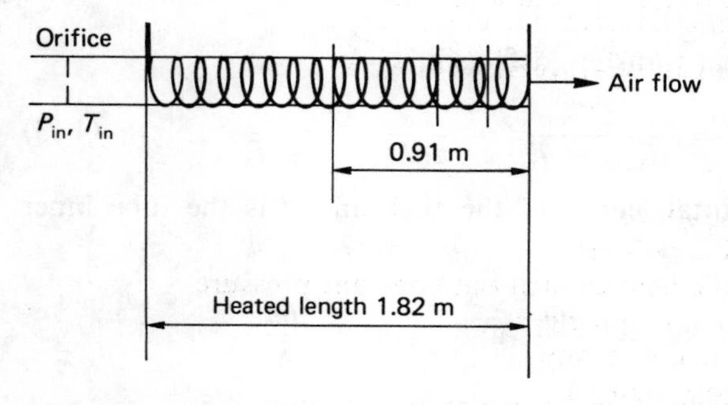

Figure 4.4

Data

Orifice pressure difference (mm H_2O)	*Orifice gauge pressure (mm H_2O)*	*Wall temperature (°C)*
110	470	38.53
92	432	39.65
67	345	41.45
36	195	45.48
16	95	52.84

Heater input	225 W	Orifice diameter	41.3 mm
Orifice inlet temperature	28 °C	Orifice coefficient of discharge	0.63
Barometer reading	101 kN/m^2	Tube inner diameter	31.76 mm

The orifice volume flow rate is given by

$$\dot{V}_{\text{orifice}} = c_d A(2gh)^{1/2} \tag{4.10}$$

The density of the air at the orifice is given by

$$\rho_{\text{orifice}} = \frac{p}{RT} \tag{4.11}$$

where p = barometer pressure + air gauge pressure at the orifice and T is the air temperature at the orifice so that the air mass flow rate is

$$\dot{m} = (\rho \dot{V})_{\text{orifice}}$$

The air temperature at the centre point measuring position is

$$T_s = T_{\text{in}} + \frac{0.5\dot{Q}}{\dot{m}c_p} \tag{4.12}$$

where T_{in} = air temperature at the orifice and $\dot{Q}$ is the heat transfer input rate.

The surface heat transfer coefficient is

$$h = \frac{\dot{Q}''}{T_{\text{wall}} - T_s} = \frac{\dot{Q}}{\pi dL(T_{\text{wall}} - T_s)} \tag{4.13}$$

where L is the total length of the tube and d is the tube inner diameter. Hence

$$\text{Re} = \frac{4\dot{m}}{\pi d\mu} \quad \text{and} \quad \text{Nu} = \frac{hd}{k}$$

The necessary reading for this program is Section 4.6.

```
10  PRINT "EXPERIMENTAL NU-RE RELATION FOR"
20  PRINT "FORCED CONVECTION IN A TUBE"
25  PRINT
30  REM  THE EXPERIMENT IS FOR TURBULENT FLOW OF AIR,PROP
ERTIES ARE REQUIRED BUT FOR SIMPLIFICATION A SINGLE SET A
T AN AVERAGE TEMPERATURE IS USED
40  REM  ESTIMATED MEAN BULK TEMPERATURE USED IS 30 C
50 VS = 0.858:CP = 1.013:K = 0.0264:MU = 0.0000185:PR = 0
.714
60  REM  SUPPLY THE EXPERIMENTAL DATA,THE CONSTANT VALUES
 IN LINE 70 AND THE VARIABLES AS READ IN DATA
70 Q = 255:T1 = 28:PA = 101:DO = 0.0413:CD = 0.63:DT = 0.
03176:L = 1.82
80  DIM P(5),HO(5),T(5),RE(5),NU(5),X(5),Y(5)
90  FOR N = 1 TO 5
100  READ P(N),HO(N),T(N)
110  NEXT N
200  REM  CALCULATION LOOP FOR RE, H AND NU
210  PRINT "REYNOLDS NUSSELT  HEAT TRANSFER"
220  PRINT " NUMBER  NUMBER  COEFFT(W/M^2 K)"
225  PRINT
230  FOR N = 1 TO 5
240 V = (3.142 * (DO ^ 2) * CD * ((2 * 9.81 * HO(N)) ^ 0.
5)) / 4
250  REM  AIR DENSITY AT ORIFICE BY GAS LAW
260 RA = (PA + (P(N) * (9.81 / 1000))) / (0.287 * (28 + 2
73))
270 M = RA * V
280  REM  FIND AIR TEMPERATURE AT MEASURING POINT USING B
ULK MEAN PROPERTIES
290 TS = T1 + (Q / (2 * M * CP * 1000))
300  REM  SURFACE HEAT TRANSFER COEFFICIENT
310 H = Q / (3.142 * DT * L * (T(N) - TS))
320 RE(N) = (4 * M) / (3.142 * DT * MU)
330 NU(N) = H * DT / K
340 RE = ( INT (RE(N) + 0.5)):NU = ( INT (NU(N) * 10 + 0.
5) / 10):H = ( INT (H * 10 + 0.5) / 10)
350  PRINT  TAB( 2)RE; TAB( 11)NU; TAB( 22)H
360  NEXT N
400  FOR Z = 1 TO 5
410 Y(Z) =  LOG ((NU(Z) / PR ^ 0.4))
420 X(Z) =  LOG (RE(Z))
430  NEXT Z
440  GOSUB 3100
450 MM =  EXP (C1):MM = ( INT (MM * 1000 + 0.5) / 1000)
460 F1 = ( INT (F1 * 100 + 0.5) / 100)
470  PRINT
480  PRINT "NU-RE RELATION FOR THIS EXPERIMENT"
490  PRINT
500  PRINT "    NU=0";MM;"(RE^0";F1;")*(PR^0.4)"
990  END
1000  DATA  470,110,38.53
1010  DATA  432,92,39.65
1020  DATA  345,67,41.45
1030  DATA  195,36,45.48
1040  DATA  95,16,52.84
3100 A1 = 0:B1 = 0:C1 = 0:D1 = 0
3110 N = 5: FOR Z = 1 TO 5
3120 A1 = A1 + X(Z)
3130 B1 = B1 + Y(Z)
3140 C1 = C1 + (X(Z) ^ 2)
3150 D1 = D1 + (X(Z) * Y(Z))
3160  NEXT Z
3170 F1 = ((N * D1) - (A1 * B1)) / ((N * C1) - (A1 ^ 2))
3180 G1 = (B1 - (F1 * A1)) / N
3200  RETURN
```

```
]RUN
EXPERIMENTAL NU-RE RELATION FOR
FORCED CONVECTION IN A TUBE

REYNOLDS NUSSELT  HEAT TRANSFER
 NUMBER  NUMBER  COEFFT(W/M^2 K)

 103871    213.7        177.6
 94658     192.6        160.1
 80124     168.1        139.8
 57904     132.3        109.9
 38235     95.4         79.3

NU-RE RELATION FOR THIS EXPERIMENT

   NU=0.026(RE^0.79)*(PR^0.4)
```

Program nomenclature

VS, CP, K, MU, PR	fluid properties as Program 4.1
Q	heat transfer rate
T1	air inlet temperature
PA	atmospheric pressure
DO	orifice diameter
CD	orifice discharge coefficient
DT	tube inner diameter
L	heated length
P	orifice pressure (gauge)
HO	orifice pressure drop
T	wall temperature at measuring station
X, Y, F1, G1	LINEFIT variables and output
V	volume flow rate at orifice
RA	density of air at orifice
M	mass flow rate of air
TS	air temperature at measuring station
H	surface heat transfer coefficient
RE	Reynolds number
NU	Nusselt number

Program notes

(1) The mean bulk temperature is estimated at 30 °C and the air properties are entered at this value. The actual temperature is determined in line 290 and will be different for each set of readings. Problem 4.3 suggests iteration of this value using Program 4.1.
(2) The program structure is as follows:

Lines 30–50 air data.
Lines 60–110 input data for problem.
Lines 200–220 print table titles.

Lines 230–310 calculation loop for h using equations (4.10) to (4.13).
Lines 320–360 calculation of Re and Nu and printing in table.
Lines 400–430 setting Re and Nu to x and y values for LINEFIT.
Lines 3100–3200 LINEFIT subroutine.
Lines 450–500 arranging LINEFIT output to Nu $= \phi(\mathrm{Re}, \mathrm{Pr})$.
Line 590 stops the program.

(3) The result shows that by assuming $\mathrm{Nu} \propto \mathrm{Pr}^{0.4}$ we obtain a relation that is reasonably close to that of equation (4.6) (but see program note 1).

Program 4.3 Surface heat transfer coefficient in smooth tubes

This program is a simple piece of software to give average surface heat transfer coefficients for forced flow in smooth tubes. As structured it uses Program 4.1 to determine property values for water flow but it could be arranged to allow for air flow (Problem 4.1).

From inputs of mass flow rate of water, tube length and diameter the program determines if flow is laminar, transitional or turbulent and if entry length or fully developed. The equations and validity ranges are shown below. Solutions will *not* be given for $2000 < \mathrm{Re} < 10\,000$ (the transition zone) nor where the flow is laminar ($\mathrm{Re} < 2000$) with the ratio $\mathrm{RePr}/(L/d) > 100$. (This ratio is often known as the Graetz number Gz.) No solutions are given for *laminar* flow situations where the thermal and velocity boundary layers start at different positions, nor if the heat transfer is by constant heat flux at the wall rather than constant wall temperature. Thus the program is limited but could, with further programming, become wider in application.

All solution equations use properties at the bulk mean temperature and give average values of Nu.

Laminar flow (entry length and fully developed) – $\mathrm{Re} < 2000$, $\mathrm{Gz} < 100$:

$$\mathrm{Nu} = 3.66 + \frac{0.0668\,\mathrm{Gz}}{1 + 0.04\,\mathrm{Gz}^{2/3}} \qquad (4.14)$$

Turbulent flow – $\mathrm{Re} > 10\,000$, $0.7 < \mathrm{Pr} < 160$:
Entry length – $L/d > 10$:

$$\mathrm{Nu} = 0.036\,\mathrm{Re}^{0.8}\,\mathrm{Pr}^{0.33}\left(\frac{d}{L}\right)^{0.055} \qquad (4.15)$$

Fully developed – $L/d > 60$:

$$\mathrm{Nu} = 0.023\ \mathrm{Re}^{0.8}\mathrm{Pr}^{0.33} \tag{4.16}$$

The necessary reading for this program is Sections 4.3, 4.4 and 4.6.

```
10   PRINT "SURFACE HEAT TRANSFER COEFFICIENT FOR"
15   PRINT "WATER FLOWING IN A SMOOTH TUBE"
20   REM  THE PROGRAM USES THE LAGRANGE INTERPOLATION FOR
WATER PROPERTIES
30   DIM T(11),VS(11),CP(11),K(11),MU(11),PR(11)
40   FOR N = 0 TO 10
50   READ T(N),VS(N),CP(N),K(N),MU(N),PR(N)
60   NEXT N
65   REM  THIS PROGRAM NEEDS AN INPUT "A" TO SET THE TEMPE
RATURE AT WHICH THE WATER PROPERTIES ARE REQUIRED...SEE L
INE1050
70   PRINT " WHAT IS THE MEAN BULK TEMPERATURE IN C?"
80   INPUT A: PRINT "WATER TEMPERATURE IS ";A;" C"
90   GOSUB 1000
100  PRINT " WHAT IS THE MASS FLOW RATE IN KG/S?"
110  INPUT M: PRINT "MASS FLOW IS ";M;" KG/S"
120  PRINT "WHAT IS THE TUBE DIAMETER IN M?"
130  INPUT D: PRINT "TUBE DIAMETER IS ";D;" M"
140  PRINT "WHAT IS THE TUBE LENGTH IN M?"
150  INPUT L: PRINT "TUBE LENGTH IS ";L;" M"
160  REM  CALCULATE THE REYNOLDS NUMBER
170 RE = 4 * M / (3.142 * D * MU)
180  PRINT "REYNOLDS NUMBER IS ";RE
190  IF RE < 10000 THEN  GOTO 300
200  PRINT " TURBULENT FLOW"
210  REM  USE THE L/D RATIO TO DETERMINE IF THE FLOW IS D
EVELOPED OR ENTRY LENGTH
220 X = L / D: PRINT "L/D IS ";X
230  IF X > 60 THEN  GOTO 270
240  PRINT "ENTRY LENGTH FLOW"
250 NU = 0.036 * (RE ^ 0.8) * (PR ^ 0.33) / (X ^ 0.055)
260  GOTO 980
270  PRINT " FULLY DEVELOPED FLOW"
280 NU = 0.023 * (RE ^ 0.8) * (PR ^ 0.33)
290  GOTO 980
300  IF RE < 2000 THEN  GOTO 400
310  PRINT "FLOW IS IN THE TRANSITION ZONE"
320  PRINT "THIS PROGRAM WILL NOT GIVE A SOLUTION"
340  END
400  PRINT "LAMINAR FLOW"
410  PRINT "THIS PROGRAM IS FOR TUBES HEATED  ALONG"
420  PRINT "THEIR WHOLE LENGTH SO THAT THE THERMAL "
430  PRINT " AND VELOCITY BOUNDARY LAYERS BOTH START AT T
HE ENTRANCE"
435  PRINT "IT IS ALSO FOR CONSTANT WALL TEMPERATURE"
440  REM  FIND THE GRAETZ NUMBER
450 X = L / D:GZ = (RE * PR) / X
455  IF GZ > 100 THEN  PRINT "GRAETZ NUMBER >100 AND ": G
OTO 320
460 NU = 3.66 + (0.0668 * GZ) / (1 + (0.04 * (GZ ^ 0.667)
))
470  GOTO 980
980 H = NU * K / D: PRINT
990  PRINT "SURFACE HEAT TRANSFER COEFFICIENT IS "H;" W/M
^2 K"
995  END
1000 B = 0:C = 0:E = 0:F = 0:G = 0
```

```
1010  FOR J = 0 TO 10
1020 X = 1
1030  FOR N = 0 TO 10
1040  IF N = J THEN  GOTO 1060
1050 X = X * (A - T(N)) / (T(J) - T(N))
1060  NEXT N
1070 B = B + X * VS(J)
1080 C = C + X * CP(J)
1090 E = E + X * K(J)
1100 F = F + X * MU(J)
1110 G = G + X * PR(J)
1115  NEXT J
1120 VS = B:CP = C:K = E:MU = F / 1000:PR = G
1130  RETURN
2000  DATA  0,0.001,4.217,0.569,1.755,13.02
2010  DATA  10,0.001,4.193,0.587,1.301,9.29
2090  DATA  90,0.00104,4.205,0.676,0.311,1.93
2100  DATA  100,0.00104,4.216,0.681,0.278,1.273
```

```
]RUN
SURFACE HEAT TRANSFER COEFFICIENT FOR
WATER FLOWING IN A SMOOTH TUBE
 WHAT IS THE MEAN BULK TEMPERATURE IN C?
?40
WATER TEMPERATURE IS 40 C
 WHAT IS THE MASS FLOW RATE IN KG/S?
?1.47
MASS FLOW IS 1.47 KG/S
WHAT IS THE TUBE DIAMETER IN M?
?.025
TUBE DIAMETER IS .025 M
WHAT IS THE TUBE LENGTH IN M?
?3
TUBE LENGTH IS 3 M
REYNOLDS NUMBER IS 114987.372
 TURBULENT FLOW
L/D IS 120
 FULLY DEVELOPED FLOW

SURFACE HEAT TRANSFER COEFFICIENT IS 10529.3733 W/M^2 K

]RUN
SURFACE HEAT TRANSFER COEFFICIENT FOR
WATER FLOWING IN A SMOOTH TUBE
 WHAT IS THE MEAN BULK TEMPERATURE IN C?
?40
WATER TEMPERATURE IS 40 C
 WHAT IS THE MASS FLOW RATE IN KG/S?
?1.47
MASS FLOW IS 1.47 KG/S
WHAT IS THE TUBE DIAMETER IN M?
?.025
TUBE DIAMETER IS .025 M
WHAT IS THE TUBE LENGTH IN M?
?1
TUBE LENGTH IS 1 M
REYNOLDS NUMBER IS 114987.372
 TURBULENT FLOW
L/D IS 40
ENTRY LENGTH FLOW

SURFACE HEAT TRANSFER COEFFICIENT IS 13454.3862 W/M^2 K

]RUN
SURFACE HEAT TRANSFER COEFFICIENT FOR
WATER FLOWING IN A SMOOTH TUBE
```

```
 WHAT IS THE MEAN BULK TEMPERATURE IN C?
?40
WATER TEMPERATURE IS 40 C
 WHAT IS THE MASS FLOW RATE IN KG/S?
?.1
MASS FLOW IS .1 KG/S
WHAT IS THE TUBE DIAMETER IN M?
?.025
TUBE DIAMETER IS .025 M
WHAT IS THE TUBE LENGTH IN M?
?3
TUBE LENGTH IS 3 M
REYNOLDS NUMBER IS 7822.2702
FLOW IS IN THE TRANSITION ZONE
THIS PROGRAM WILL NOT GIVE A SOLUTION
]RUN
SURFACE HEAT TRANSFER COEFFICIENT FOR
WATER FLOWING IN A SMOOTH TUBE
 WHAT IS THE MEAN BULK TEMPERATURE IN C?
?40
WATER TEMPERATURE IS 40 C
 WHAT IS THE MASS FLOW RATE IN KG/S?
?.0147
MASS FLOW IS .0147 KG/S
WHAT IS THE TUBE DIAMETER IN M?
?.025
TUBE DIAMETER IS .025 M
WHAT IS THE TUBE LENGTH IN M?
?3
TUBE LENGTH IS 3 M
REYNOLDS NUMBER IS 1149.87372
LAMINAR FLOW
THIS PROGRAM IS FOR TUBES HEATED  ALONG
THEIR WHOLE LENGTH SO THAT THE THERMAL
 AND VELOCITY BOUNDARY LAYERS BOTH START AT THE ENTRANCE
IT IS ALSO FOR CONSTANT WALL TEMPERATURE

SURFACE HEAT TRANSFER COEFFICIENT IS 139.69566 W/M^2 K

]RUN
SURFACE HEAT TRANSFER COEFFICIENT FOR
WATER FLOWING IN A SMOOTH TUBE
 WHAT IS THE MEAN BULK TEMPERATURE IN C?
?40
WATER TEMPERATURE IS 40 C
 WHAT IS THE MASS FLOW RATE IN KG/S?
?.0147
MASS FLOW IS .0147 KG/S
WHAT IS THE TUBE DIAMETER IN M?
?.025
TUBE DIAMETER IS .025 M
WHAT IS THE TUBE LENGTH IN M?
?.1
TUBE LENGTH IS .1 M
REYNOLDS NUMBER IS 1149.87372
LAMINAR FLOW
THIS PROGRAM IS FOR TUBES HEATED  ALONG
THEIR WHOLE LENGTH SO THAT THE THERMAL
 AND VELOCITY BOUNDARY LAYERS BOTH START AT THE ENTRANCE
IT IS ALSO FOR CONSTANT WALL TEMPERATURE
GRAETZ NUMBER >100 AND
THIS PROGRAM WILL NOT GIVE A SOLUTION
```

Program nomenclature

T, VS, CP, K, MU, PR	fluid properties in Program 4.1
A	the bulk mean temperature
M	mass flow rate
D	tube diameter
L	tube length
RE	Reynolds number
NU	Nusselt number
PR	Prandtl number
GZ	Graetz number
H	surface heat transfer coefficient

Program notes

(1) The program structure is as follows:

Lines 30–90 and 1000–2100 Program 4.1 for properties.
Lines 100–150 input data M, D, L.
Line 190 go to laminar flow or transitional flow.
Lines 210–290 turbulent flow solutions.
Line 300 go to laminar flow.
Lines 310–330 print no solution if program invalid.
Lines 400–470 laminar flow solutions.
Lines 980–995 determine and print surface heat transfer coefficient.

(2) Five runs are shown covering three solutions (turbulent: entry length and fully developed, laminar) and two invalid cases (transition zone and laminar flow with Gz > 100).

Program 4.4 Free convection at vertical surfaces

Write a program to find the surface heat transfer coefficient in free convection at a plane vertical surface at 80 °C in air at 20 °C. Consider surface heights from 0.1 m to 2 m in increments of 0.1 m.

This program is similar to Program 4.3 but determines average surface heat transfer coefficients in free convection at vertical plane *or cylindrical* surfaces. From the height of the surface the program decides if flow is laminar or turbulent and applies equations (4.9):

$$\left.\begin{aligned} &\text{Laminar Ra} < 10^9: && \text{Nu} = 0.59\,(\text{Pr}\cdot\text{Gr})^{1/4} = 0.59\,\text{Ra}^{1/4} \\ &\text{Turbulent Ra} > 10^9: && \text{Nu} = 0.13\,(\text{Pr}\cdot\text{Gr})^{1/3} = 0.13\,\text{Ra}^{1/3} \end{aligned}\right\} \quad (4.9)$$

Examination of these relations shows that in *turbulent* flow:

$$\text{Nu} \propto (\text{Gr}\cdot\text{Pr})^{1/3} \quad \text{but} \quad \text{r} = \frac{\rho^2 \beta g \theta l^3}{\mu^2}$$

where l is surface height. Thus

$$\text{Nu} \propto \text{Pr}\cdot l$$

But

$$h = \frac{\text{Nu}\cdot k}{l} \quad \text{so that} \quad h \propto \text{Pr}^{1/3}\cdot k$$

Thus, for constant film temperature, h = constant. We can therefore expect the turbulent program to give a constant h.

For *laminar* flow

$$h \propto \text{Pr}^{1/4}\cdot k \cdot l^{-1/4}$$

so that we can expect h to decrease as height increases with constant film temperature. The necessary reading for this program is Section 4.7.

```
10  PRINT "FREE CONVECTION AT VERTICAL SURFACES"
15  PRINT
20  REM  THIS PROGRAM INVESTIGATES THE EFFECT OF SURFACE
HEIGHT ON HEAT TRANSFER COEFFICIENT
30  PRINT "HEIGHT   RAYLEIGH  SURFACE   HEAT   NUSSELT
35  PRINT "          NUMBER   COEFFT. TRANSFER NUMBER"
37  PRINT "   M      X 10^4   W/M^2 K       W"
38  PRINT
39  DIM H(20)
40  FOR N = 1 TO 20
50  READ H(N)
60  NEXT N
70  REM  AIR PRERTIES AT THE FILM TEMPERATURE FROM STATEM
ENT(OR INTERPOLATION ROUTINE)
80  REM  FILM TEMPERATURE IS 50 C
90 VS = 0.915:CP = 1.015:K = 0.028:MU = 0.0198:PR = 0.71
100  REM  COEFFICIENT OF CUBICAL EXPANSION IS ALSO REQUIR
ED:FOR A PERFECT GAS B=1/T
110 B = (1 / 323):TS = 80:TA = 20
120  FOR N = 1 TO 20
130 RA = (((1 / VS) ^ 2) * (H(N) ^ 3) * CP * B * 9.81 * (
TS - TA) * 1000) / ((K * MU) / 1000)
140  IF RA >  = 1000000000 THEN  GOTO 170
150 NU = 0.59 * (RA ^ 0.25)
160  GOTO 180
170 NU = 0.13 * (RA ^ 0.33)
180 HTC = NU * K / H(N)
190 Q = HTC * 1.6 * (TS - TA)
200 RA =  INT (RA) / 10000:HTC =  INT (HTC * 100 + 0.5) /
 100:Q =  INT (Q * 10 + 0.5) / 10:NU =  INT (NU * 10 + 0.
5) / 10
220  PRINT  TAB( 4)H(N); TAB( 9)RA; TAB( 21)HTC; TAB( 29)
Q; TAB( 37)NU
230  NEXT N
```

```
300  DATA  0.1,0.2,0.3,0.4,0.5,0.6,0.7,0.8,0.9,1.0,1.1,1.
2,1.3,1.4,1.5,1.6,1.7,1.8,1.9,2.0

]RUN
FREE CONVECTION AT VERTICAL SURFACES

HEIGHT   RAYLEIGH  SURFACE   HEAT    NUSSELT
          NUMBER   COEFFT.  TRANSFER NUMBER
  M       X 10^4   W/M^2 K     W

  .1    398.4908     7.38    708.6    26.4
  .2    3187.9269    6.21    595.8    44.3
  .3    10759.2533   5.61    538.4    60.1
  .4    25503.4153   5.22    501      74.6
  .5    49811.3579   4.94    473.9    88.1
  .6    86074.0265   4.72    452.7    101.1
  .7    136682.366   5.38    516.5    134.5
  .8    204027.322   5.37    515.8    153.5
  .9    290499.839   5.37    515.2    172.5
  1     398490.864   5.36    514.6    191.5
  1.1   530391.34    5.36    514.2    210.4
  1.2   688592.213   5.35    513.7    229.3
  1.3   875484.428   5.35    513.3    248.2
  1.4   1093458.93   5.34    512.9    267.1
  1.5   1344906.66   5.34    512.6    286
  1.6   1632218.58   5.34    512.2    304.9
  1.7   1957785.62   5.33    511.9    323.8
  1.8   2323998.72   5.33    511.6    342.6
  1.9   2733248.84   5.33    511.4    361.4
  2     3187926.91   5.32    511.1    380.3
```

Program nomenclature

H	surface height
VS, CP, K, MU, PR	fluid properties
B	coefficient of cubical expansion
TS	surface temperature
TA	air temperature
RA	Rayleigh number
NU	Nusselt number
HTC	surface heat transfer coefficient
Q	heat transfer rate

Program notes

(1) The program structure is simple. The only crucial step is line 140 which chooses laminar or turbulent flow.
(2) In the print out arrangements the numbers have been rounded off and Ra is reduced by a factor 10^4 for convenience.
(3) The change to turbulent flow occurs between heights of 0.6 m and 0.7 m, thus for laminar flow between 0.1 m and 0.6 m:

$$h \propto \text{(height)}^{-1/4}$$

we find

$$\frac{h_{0.1}}{h_{0.6}} = 1.564 \quad \text{and} \quad \left(\frac{0.1}{0.6}\right)^{-1/4} = 1.565$$

as expected.

For turbulent flow h should be constant but it appears to decrease. This is due to a *programming error* in using 0.33 for $\frac{1}{3}$ in line 170. A hand calculation demonstrates that for $l = 1$

$$\text{Ra} = 3\,984\,908\,640 \quad \text{giving} \quad h = 5.771\,\text{W/m}^2\,\text{K}$$

and for $l = 2$

$$\text{Ra} = 31\,879\,369\,100 \quad \text{giving} \quad h = 5.771\,\text{W/m}^2\,\text{K}$$

It is important to realise that approximations may give misleading results.

Program 4.5 Laminar film condensation

When a vapour condenses continuously on an inclined surface and the condensed liquid forms a film on the surface as it flows off under gravity, the process is known as *film condensation*. Analysis of laminar film condensation gives, for inclined plates or tubes [1],

$$h = 0.943\left[\frac{g\rho_l(\rho_l - \rho_v)h_{fg}k_l^3 \sin\theta}{\mu_l(T_v - T_w)L}\right]^{1/4} \tag{4.17}$$

Similarly, for a horizontal tube it is found that in laminar film condensation

$$h = 0.725\left[\frac{g\rho_l(\rho_l - \rho_v)h_{fg}k_l^3}{\mu_l(T_v - T_w)d}\right]^{1/4} \tag{4.18}$$

where h is the *average* heat transfer coefficient
g is gravitational acceleration
ρ_l is the liquid density
ρ_v is the vapour density
h_{fg} is the enthalpy of vaporisation
k_l is the liquid thermal conductivity
θ is the plate inclination to the horizontal
μ_l is the liquid dynamic viscosity
T_v is saturation temperature of the vapour
T_w is the wall temperature
L is the plate length
and d is the tube diameter

For vertical tiers of N horizontal tubes, each of diameter d,

$$h = \frac{1}{N^{1/4}} \cdot h_{\text{horizontal tube}}$$

Properties for use in equations (4.17) and (4.18) should be evaluated as the film temperature

$$T_{\text{film}} = \tfrac{1}{2}(T_{\text{wall}} + T_{\text{saturation}})$$

Write a program to give film condensation heat transfer coefficients for vertical and inclined surfaces and for single cases and tiers of horizontal tubes; hence solve the following problem.

Saturated steam at 85 kN/m^2 condenses on a surface at 25 °C which is:

(a) a vertical plate 3 m in length;
(b) an inclined plate 3 m in length at 60° to the horizontal;
(c) a single horizontal tube 0.015 m in diameter;
(d) an array of 100 tubes in a 10 × 10 vertical matrix each tube being 0.015 m in diameter.

The necessary reading for this program is Section 4.8.

At 85 kN/m^2:

$$T_{\text{saturation}} = 95.2\,°\text{C} \quad \text{and} \quad T_{\text{wall}} = 25\,°\text{C}$$

Thus

$$T_{\text{film}} = \left(\frac{25 + 95.2}{2}\right) = 60.1\,°\text{C}$$

Reference

1. Ozisik, M.D. *Heat Transfer*, McGraw Hill (1985).

```
10  PRINT "LAMINAR FILM CONDENSATION"
20  PRINT "ON PLANE VERTICAL OR INCLINED SURFACES"
30  PRINT "AND FOR HORIZONTAL TUBES AND VERTICAL TIERS "
40  PRINT " OF HORIZONTAL TUBES"
50  REM  PROPERTIES ARE REQUIRED  AT THE FILM TEMPERATURE

60  PRINT "WHAT IS THE CONDENSATE DENSITY IN KG/M^3?"
70  INPUT RL: PRINT "CONDENSATE DENSITY IS "RL;" KG/M^3"
80  PRINT "WHAT IS THE VAPOUR DENSITY IN KG/M^3?"
90  INPUT RV: PRINT "VAPOUR DENSITY IS ";RV;" KG/M^3"
100  PRINT "WHAT IS THE SPECIFIC ENTHALPY OF VAPORISATION
 IN KJ/KG K?"
110  INPUT HFG: PRINT "ENTHALPY OF VAPORISATION IS ";HFG;
" KJ/KG K"
120  PRINT "WHAT IS THE THERMAL CONDUCTIVITY OF THE CONDE
NSATE IN W/M K?"
130  INPUT KL: PRINT "THERMAL CONDUCTIVITY OF CONDENSATE
IS ";KL;" W/M K"
140  PRINT "WHAT IS THE DYNAMIC VISCOSITY OF THE CONDENSA
TE IN KG/M S?"
```

```
150  INPUT MUL: PRINT "CONDENSATE VISCOSITY IS ";MUL;" KG
/M S"
160  PRINT "WHAT IS THE SATURATION TEMPERATURE IN C?"
170  INPUT TV: PRINT "SATURATION TEMPERATURE IS ";TV;" C"

180  PRINT "WHAT IS THE SURFACE TEMPERATURE IN C?"
190  INPUT TW: PRINT "SURFACE TEMPERATURE IS ";TW;" C"
200 Z = ((9.81 * RL * (RL - RV) * HFG * 1000 * (KL ^ 3))
/ (MUL * (TV - TW))) ^ 0.25
210  PRINT "IS THE SURFACE A PLATE OR A TUBE?"
215  PRINT "TYPE 1 FOR A TUBE OR 9 FOR A PLATE"
220  INPUT X: IF X = 9 THEN  GOTO 290
230  PRINT "WHAT IS THE TUBE DIAMETER IN M?"
240  INPUT D
250  PRINT " HOW MANY TUBES ARE THERE IN EACH TIER?"
260  INPUT N: PRINT " THERE ARE ";N;" TUBES OF DIAMETER "
;D;" M IN EACH TIER"
270 H = (Z * 0.725) / ((N * D) ^ 0.25)
280  GOTO 400
290  PRINT "WHAT IS THE PLATE LENGTH IN M?"
300  INPUT L: PRINT "PLATE LENGTH IS ";L;" M"
310  PRINT "WHAT IS THE PLATE INCLINATION FROM THE HORIZO
NTAL IN DEGREES?"
320  INPUT A: PRINT " PLATE INCLINATION IS ";A;" DEGREES"

330 A = (A * 3.142) / 180
340 H = (Z * 0.943) * (( SIN (A) / L) ^ 0.25)
400  PRINT "SURFACE HEAT TRANSFER COEFFICIENT IS ";H;" W/
M^2 K"
```

```
]RUN
LAMINAR FILM CONDENSATION
ON PLANE VERTICAL OR INCLINED SURFACES
AND FOR HORIZONTAL TUBES AND VERTICAL TIERS
 OF HORIZONTAL TUBES
WHAT IS THE CONDENSATE DENSITY IN KG/M^3?
?983.3
CONDENSATE DENSITY IS 983.3 KG/M^3
WHAT IS THE VAPOUR DENSITY IN KG/M^3?
?0.131
VAPOUR DENSITY IS .131 KG/M^3
WHAT IS THE SPECIFIC ENTHALPY OF VAPORISATION IN KJ/KG K?
?2358.4
ENTHALPY OF VAPORISATION IS 2358.4 KJ/KG K
WHAT IS THE THERMAL CONDUCTIVITY OF THE CONDENSATE IN W/M K?
?0.653
THERMAL CONDUCTIVITY OF CONDENSATE IS .653 W/M K
WHAT IS THE DYNAMIC VISCOSITY OF THE CONDENSATE IN KG/M S?
?0.000462
CONDENSATE VISCOSITY IS 4.62E-04 KG/M S
WHAT IS THE SATURATION TEMPERATURE IN C?
?95.2
SATURATION TEMPERATURE IS 95.2 C
WHAT IS THE SURFACE TEMPERATURE IN C?
?25
SURFACE TEMPERATURE IS 25 C
IS THE SURFACE A PLATE OR A TUBE?
TYPE 1 FOR A TUBE OR 9 FOR A PLATE
?1
WHAT IS THE TUBE DIAMETER IN M?
?0.015
 HOW MANY TUBES ARE THERE IN EACH TIER?
?10
 THERE ARE 10 TUBES OF DIAMETER .015 M IN EACH TIER
SURFACE HEAT TRANSFER COEFFICIENT IS 4336.66104 W/M^2 K
```

Program nomenclature

RL liquid density
RV vapour density
HFG enthalpy of vaporisation
KL liquid thermal conductivity
MUL liquid dynamic viscosity
TV vapour saturation temperature
TW wall temperature
Z common function to all equations
H surface heat transfer coefficient
L plate length
D tube diameter
N number of tubes per tier
A plate inclination

Program notes

(1) The program structure is simple, as follows:

Lines 50–150 are data inputs, the film temperature being 60.1 °C.
Lines 160–190 are further data inputs.
Line 200 calculates the common parameter Z.
Line 210 chooses plate or tube.
Lines 230–270 solve tube problems.
Lines 290–340 solve plate problems.

(2) Only the run for the 100 tube array is shown, the answers for the four parts being (a) 2667.3 W/m^2 K, (b) 2573.1 W/m^2 K, (c) 7771.8 W/m^2 K, (d) 4336.7 W/m^2 K, which clearly illustrate the high heat transfer coefficients that are achieved in condensation (phase change) heat transfer.
(3) Experiments for condensation on *vertical* surfaces suggest that h is underestimated by about 20% by equation (4.17) and the constant is quoted as 1.13 in some references. (The original analysis equation (4.17) was derived by Nusselt in 1916.)
(4) To allow for subcooling of the condensate film some references suggest that h_{fg} is modified to h'_{fg} where

$$h'_{fg} = h_{fg} + \tfrac{3}{8} c_{p_l} (T_{saturation} - T_{wall})$$

When this method is used it is often found that the fluid properties are not evaluated at the film temperature but at some other prescribed value.

PROBLEMS

(4.1) Rearrange Program 4.1 to incorporate the air data as well as the water data. Lines 30–100 can be arranged to read either water or air (or data for any other substance; refrigerant 12 might be useful). Having done this, rearrange Programs 4.4 and 4.5 to use the program. Program 4.5 will also need a special block over a suitable range of condensing temperatures for steam.

(4.2) In an experiment to determine a forced convection surface heat transfer coefficient relationship for a cylinder in cross-flow the results shown below were obtained. In the experiment cold air was blown at various flow rates over a cylinder repeatedly heated to the same temperature T_{max} and then allowed to cool to another fixed temperature T_{min} in a measured time. The cylinder may be treated as a lumped capacity system. The air flow rate is determined by a pitot static tube for which

$$V = (2gh)^{1/2}$$

where V is the air velocity, g is 9.81 m/s^2 and h is the pitot static head in metres of air.

Write a program to determine the relationship between Reynolds number and Nusselt number using the cylinder diameter as the characteristic length dimension. Properties should be determined at the free stream temperature. The LINEFIT program may be used to obtain the relationship:

Time to cool from T_{max} to T_{min} (seconds)	*Pitot static head (mm water)*
117	40.5
128	33.3
143	23.5
162	14.2
190	7.4
218	4.3
270	1.7

Cylinder maximum temperature	63 °C
Cylinder minimum temperature	19 °C
Air temperature	14 °C
Mass of cylinder	0.1067 kg
Surface area of cylinder	0.004 05 m^2
Cylinder diameter	0.012 46 m
Specific heat capacity of cylinder	380 J/kg K

(4.3) Program 4.3 used an estimated bulk mean temperature which was never checked to see if it was correct. Rearrange the program to check this estimate and iterate the estimate asking for a new set of properties each time or by including Program 4.1. Does this step affect the results significantly? Try to find how far the initial estimate has to be in error to cause a significant error in the result.
(4.4) Write a program to compare the surface heat transfer coefficient for forced, turbulent flow in smooth tubes over a suitable range of Reynolds number determined by:

(a) the simple Reynolds analogy

$$\mathrm{St} = \frac{f}{2}$$

(b) the Colburn modification

$$\mathrm{St}\,\mathrm{Pr}^{2/3} = \frac{f}{2}$$

(c) the Prandtl modification

$$\mathrm{St} = \frac{f}{2}\left[\frac{1}{1 + 5\sqrt{\frac{f}{2}}\,(\mathrm{Pr} - 1)}\right]$$

(d) the Prandtl–Taylor modification

$$\mathrm{St} = \frac{f}{2}\left[\frac{1}{1 + r_{\mathrm{v}}(\mathrm{Pr} - 1)}\right]$$

where $r_{\mathrm{v}} = 2.11\,\mathrm{Re}^{-0.1}$.

Consider water flowing in a 25 mm diameter tube with Reynolds number changing from 10 000 to 100 000 in 10 000 steps and mean bulk temperatures of 10–100 °C in 10 K steps.
(4.5) Write a program for flat plate surface heat transfer coefficients in forced flow similar to Program 4.3.
(4.6) Write programs for other free convection situations such as horizontal surfaces, inclined surfaces, vertical 'double glazed' spaces and combine them with Program 4.4 as a piece of software.
(4.7) Consider the problem of turbulence in film condensation on vertical surfaces which may occur towards the lower end of a long surface. Program 4.5 *assumes* laminar flow and, as it stands, may be invalid in some data input situations. It is at present of dubious value until a critical Reynolds number is programmed. This is important because in turbulent condensate flow the heat transfer coefficients

increase with length, whereas they have clearly decreased with length in laminar flow (equation (4.17)). The value of the critical Reynolds number is about 1800 and Re is defined by

$$\mathrm{Re} = \frac{4\dot{m}}{\mu_l P}$$

where $\dot{m}$ is the mass flow rate of condensate ($\dot{m}$ will increase down the surface)
μ_l is the dynamic viscosity of the liquid at the film temperature
and P is the wetted perimeter (πD for a tube and W the width for a vertical or inclined plate)

Clearly, this will be a calculation where h is obtained from the laminar relation and $\dot{m}$ is determined at the *lowest* point on the surface and hence is Re determined. If Re > 1800 the relation for the average surface heat transfer coefficient is changed to

$$h = 0.0076\ \mathrm{Re}^{0.4}\left(\frac{k_l^3 \rho_l^2 g}{\mu_l^2}\right)^{1/3} \qquad (4.19)$$

where suffix l refers to liquid properties determined at

$$\frac{T_{\mathrm{wall}} + T_{\mathrm{saturation}}}{2}$$

Write the necessary program to ensure Program 4.5 is not in error and to determine surface heat transfer coefficients for surfaces when Re > 1800.

(4.8) In all the worked programs using Nu = ϕ(Re, Pr) for flow in tubes, simple relations have been used which can give rise to inaccuracy (Section 4.10). Reference 4 suggests that the use of the Petukhov relation gives better results for the common case of forced flow in tubes which may be rough or smooth. The programs to which this could be applied include those with fully developed turbulent flow: 4.3. 7.1, 7.2 and 7.4. The relation is

$$\mathrm{Nu} = \frac{\mathrm{Re\,Pr}}{X}\left(\frac{f}{2}\right)\left(\frac{\mu_b}{\mu_w}\right)^n$$

where

$$X = 1.07 + 12.7(\mathrm{Pr}^{2/3} - 1)\left(\frac{f}{2}\right)^{1/2}$$

and

$$f = \tfrac{1}{4}(1.82 \log_{10} \mathrm{Re} - 1.64)^{-2}$$

Also,

$n = 0.11$ for heating with uniform wall temperature
$n = 0.25$ for cooling with uniform wall temperature
$n = 0$ for uniform wall heat flux analysis

Validity ranges are

$0.5 < \mathrm{Pr} < 200$ with 5–6% error
$0.5 < \mathrm{Pr} < 2000$ with 10% error
$10^4 < \mathrm{Re} < 5 \times 10^6$
$0.08 < \mu_w/\mu_b < 40$

Properties (except μ_w) are evaluated at the mean bulk temperature and μ_w is evaluated at the wall temperature.

(4.9) A glasshouse is heated by tubular heaters fitted horizontally. The length of heater is L per m^2 of glass surface area. The tube diameter is D. The energy for heating the glasshouse is supplied by water which has been used as the condenser coolant in an adjacent steam power station. This water gives a tube surface temperature of 28 °C. The minimum outside temperature expected is −10 °C and the glasshouse is to be kept at a minimum temperature of 17 °C. The U value of the glazing is 6 W/m² K. Write a program to determine L for various heating tube diameters between 30 and 120 mm. (Problem 4.6 should have included a suitable method for the surface heat transfer coefficient of the tubes.) Extend the program to allow for the design of a control system which will need to operate when the outside temperature is greater than −10 °C.

(4.10) Heat flux through the cylinder walls of a reciprocating internal combustion engine may be determined using the Eichelberg relation:

$$h = 2.1c^{1/3}(pT)^{1/2}\ \mathrm{kcal/m^2\,hour\,K}$$

where c is the mean piston speed in m/s
p is the instantaneous cylinder pressure in atmospheres
and T is the instantaneous gas temperature in K

Write a program to determine the surface heat transfer coefficient in W/m² K and the heat transfer in joules per degree of crank angle for any crank angle θ. Evaluate the program for an engine with bore 0.0762 m, stroke 0.1111 m with connecting rod length 0.2413 m running at 1750 rev/min with the following data:

Crank angle (°)	*Pressure* (kN/m^2)	*Temperature* (*K*)
220	85	300
300	313	530
360	4400	1020
390	4700	2800
450	320	900

The area for heat transfer is determined from the piston position given by

$$x = l + r - [r \cos \theta + (l^2 - r^2 \sin^2 \theta)^{1/2}]$$

where x is the piston displacement from top dead centre, r is the crank length, l is the connecting rod length and θ is the crank angle measuring from the top dead centre position in a clockwise direction. The surface temperature of the cylinder walls may be assumed constant at 380 K.

Open-ended problems

(**4.11**) Evaporation in pool boiling (as opposed to boiling in flow in tubes) exhibits the characteristics shown in Figure 4.5. Make a study

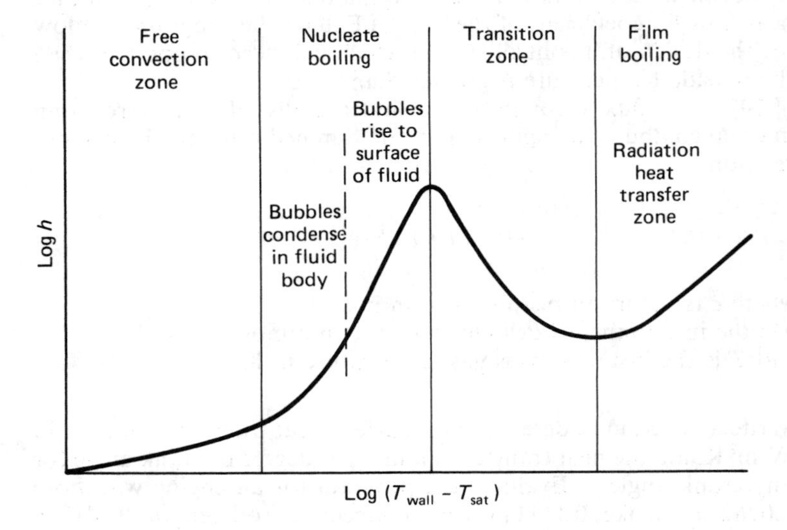

Figure 4.5

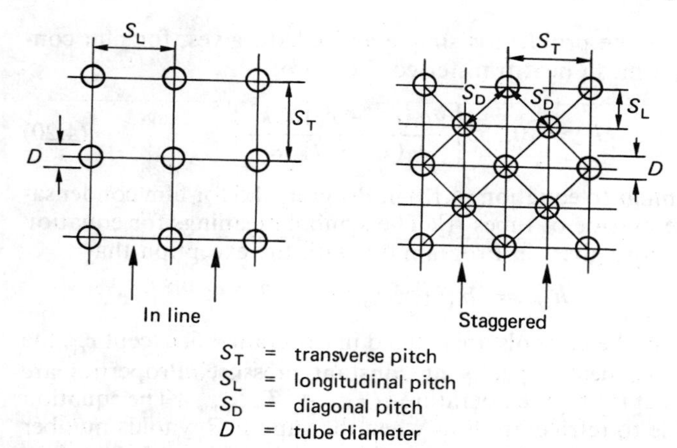

Figure 4.6

of the zones and write a program to enable the surface heat transfer coefficients to be determined. The work will involve simple free convection from a heated surface, nucleate boiling by the Rohsenow relation and film boiling by the Bromley relation, together with Kutateladze or Zuber relations for peak heat flux. Information may be found in the references to Chapter 4 and elsewhere.

(4.12) Heat exchangers often have tubes in banks or bundles characterised by their pitch (Figure 4.6). For such arrangements, the Reynolds numbers are based on the velocity of flow determined at the minimum free flow area available in the bundle. When this Reynolds number has been determined it is found that

$$\text{Nu} = \phi(\text{Re, Pr, pitch})$$

and there are correlations to determine surface heat transfer coefficients notably by Grimison or Zukauskas, both of which involve the use of tables of constants and indices. These tables are available in the references to Chapter 4 and elsewhere. Write a program to determine the heat transfer coefficients and the pressure drops in tube banks suitable for use in the design of shell and tube heat exchangers.

(4.13) Program 4.5 determined surface heat transfer coefficients for condensation on plates or the *outside* of tubes. In practice, condensation often occurs inside tubes, in which case there is a flow along the tubes which will involve the movement of both liquid and vapour. This complicates the problem and, in particular for vertical tubes, the flow of vapour may be upward or downward whereas free liquid flow will always be downward due to gravitational effects. In horizontal tubes the problem is simple and Chato gives, for film condensation, a mean heat transfer coefficient of

$$h = 0.555\left[\frac{g\rho_l(\rho_l - \rho_v)h'_{fg}k_k^3}{\mu_l(T_v - T_w)d}\right]^{1/4} \qquad (4.20)$$

which is similar to equation (4.18) in Program 4.5 for film condensation on the outside of tubes [1]. The symbol meanings for equation (4.20) are those given in Program 4.5 with the exception that

$$h'_{fg} = h_{fg} + \tfrac{3}{8}c_{pl}(T_v - T_w)$$

where, again, the symbols are defined in Program 4.5, except c_{p_l}, the liquid specific heat capacity at constant pressure. Properties are determined at the film temperature $\frac{1}{2}(T_{wall} + T_{saturation})$. The equation is applicable to refrigerant flow when the vapour Reynolds number Re_v, which must be evaluated at the inlet conditions, is less than 35 000:

$$Re_v = \frac{\rho_v V_v d}{\mu_v}$$

A refrigerator condenser takes superheated vapour and cools it to the saturation temperature, then condenses it and subcools the liquid. Consider refrigerant 12 (see Problem 4.1) at a pressure of 0.745 MN/m^2 entering a tube of diameter d mm at a temperature of 50 °C and leaving the tube at 25 °C. The tube is arranged to be mainly horizontal by a zig–zag layout on the back of the refrigerator (examine the back of a domestic vapour compression refrigerator). Vary the mass flow rate to keep within the Reynolds number criterion in the Chato relation and choose tube diameters of the order found on the back of domestic refrigerators; hence make an estimate of the cooling capacity of such a machine. Compare this with the maker's rating.

Reference

1. Chato, J.C. Laminar condensation inside horizontal and inclined tubes, *Journal of the American Society of Heating, Refrigeration and Airconditioning Engineers*, **4,** 52 (1962).

Chapter 5

Radiation

5.1 Introduction

Simple radiation concepts were included in Chapter 2, enabling solutions involving black or grey bodies in large enclosures. In order to solve more general problems it is necessary to know the directional distribution of energy from a radiating surface and the amount of energy that will be intercepted by any non-enclosing surface that can 'see' the radiating surface.

5.2 Intensity of radiation

Consider a small area δA of emissive power $\dot{E}''$. The emitted energy

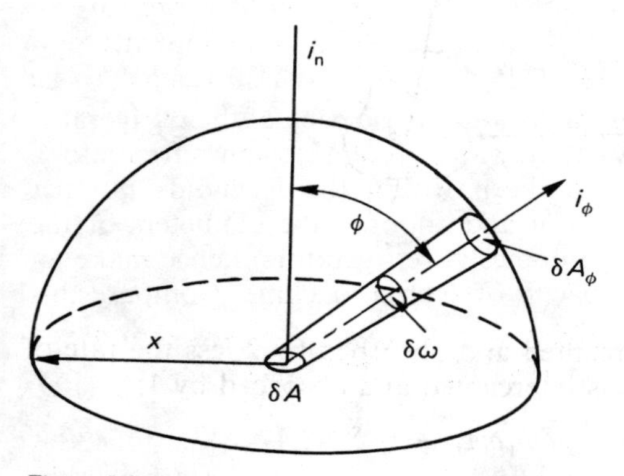

Figure 5.1

must all pass through the hemispherical surface shown in Figure 5.1. The intensity of radiation i_ϕ in direction ϕ is defined as

$$i_\phi = \left(\frac{\delta \dot{E}''}{\delta \omega}\right)_\phi \tag{5.1}$$

where $\delta\omega$ is the solid angle subtended by the area δA_ϕ at the emitting area δA:

$$\delta\omega = \left(\frac{\delta A_\phi}{x^2}\right) \tag{5.2}$$

Lambert's law of diffuse radiation states that

$$i_\phi = i_n \cos \phi \tag{5.3}$$

where i_n is the normal intensity of radiation. For black surfaces:

$$i_n = \frac{\sigma T^4}{\pi} \tag{5.4}$$

For grey surfaces:

$$i_n = \frac{\varepsilon \sigma T^4}{\pi} \tag{5.5}$$

5.3 The geometric factor

Consider the two small black surfaces shown in Figure 5.2. The heat transfer rate from surface 1 to surface 2 is given by the rate of energy

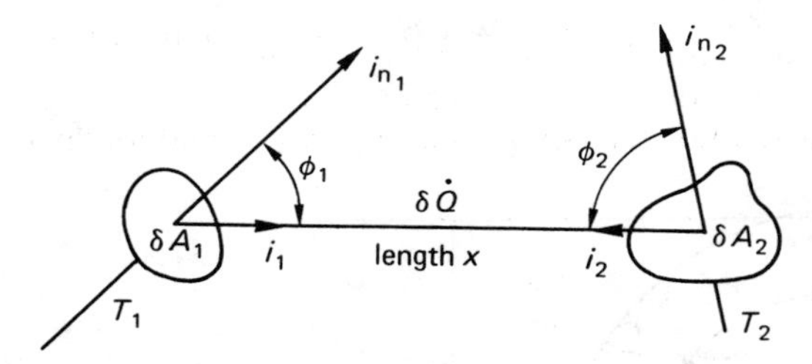

Figure 5.2

emitted by 1 that is intercepted and absorbed by 2 less the rate of energy emitted by 2 that is intercepted and absorbed by 1:

$$\delta\dot{Q}_{12} = i_{\phi_1}\delta\omega_1\delta A_1 - i_{\phi_2}\delta\omega_2\delta A_2$$

where

$$\delta\omega_1 = \frac{\delta A_2 \cos \phi_2}{x^2} \quad \text{and} \quad \delta\omega_2 = \frac{\delta A_1 \cos \phi_1}{x^2}$$

Hence using equations (5.3) and (5.4)

$$\delta\dot{Q}_{12} = \sigma(T_1^4 - T_2^4)\left[\frac{\cos \phi_1 \cos \phi_2 \delta A_1 \delta A_2}{\pi x^2}\right]$$

For a finite black area this becomes

$$\dot{Q}_{12} = \sigma(T_1^4 - T_2^4) \int_{A_1}\int_{A_2} \frac{\cos\phi_1 \cos\phi_2 \, dA_1 \, dA_2}{\pi x^2} \quad (5.6)$$

A geometric factor F is defined as the fraction of the energy emitted per unit area by one surface that is intercepted by another surface. Hence

$$F_{12} = \frac{1}{A_1} \int_{A_1}\int_{A_2} \frac{\cos\phi_1 \cos\phi_2 \, dA_1 \, dA_2}{\pi x^2}$$

and

$$F_{21} = \frac{1}{A_2} \int_{A_1}\int_{A_2} \frac{\cos\phi_1 \cos\phi_2 \, dA_1 \, dA_2}{\pi x^2} \quad (5.7)$$

and

$$A_1 F_{12} = A_2 F_{21} \quad (5.8)$$

From equations (5.6) and (5.8)

$$\dot{Q}_{12} = A_1 F_{12} \sigma(T_1^4 - T_2^4) = A_2 F_{21} \sigma(T_1^4 - T_2^4) \quad (5.9)$$

Determination of geometric factors is often tedious and data is available [1] by use of equations (Program 5.2) or charts (Figure 5.3). Geometric factors (also known as shape factors) can be manipulated

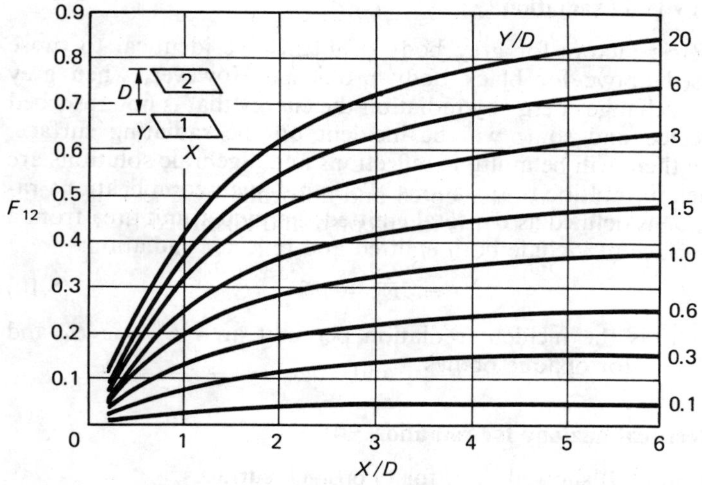

Geometric factor for two rectangular parallel plates

Figure 5.3

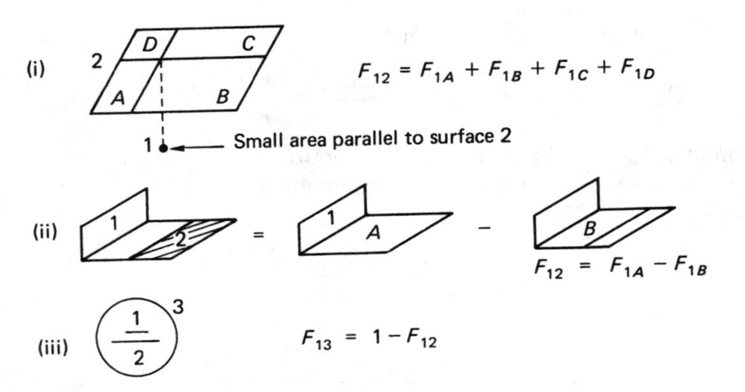

Figure 5.4

with *shape factor algebra* to produce information for other geometries but care is needed to avoid errors [2]. Simple examples are shown in Figure 5.4. Equation (5.8) and Figure 5.4 indicate that a considerable amount of information can be obtained from a single chart. In general for an enclosure involving N surfaces there will be N^2 geometric factors but only $N(N - 1)/2$ of these will be independent. Geometric factors for N surface enclosures are often arranged in matrix form [3].

5.4 Grey body radiation

Geometric factors for grey body problems are identical to those discussed above for black body problems. However, when grey bodies exchange energy by radiation the energy that is not absorbed is reflected and some will be incident on the radiating surface. Clearly there will be multiple reflections and algebraic solutions are complex in multibody situations. To avoid such complications radiosity $\dot{J}''$ is defined as the total emitted energy per unit area from a grey surface to include both emitted and reflected radiation:

$$\dot{J}'' = \dot{E}''_g + \rho \dot{G}'' \tag{5.10}$$

where $\dot{G}''$ is the incident radiation per unit area $\dot{E}''_g = \varepsilon \cdot \dot{E}''_b$ and $\rho = 1 - \varepsilon$ for opaque bodies.

5.5 Electrical analogy for radiation

Equation (5.9) showed that, for two black surfaces,

$$\dot{Q}_{12} = A_1 F_{12} \sigma (T_1^4 - T_2^4)$$

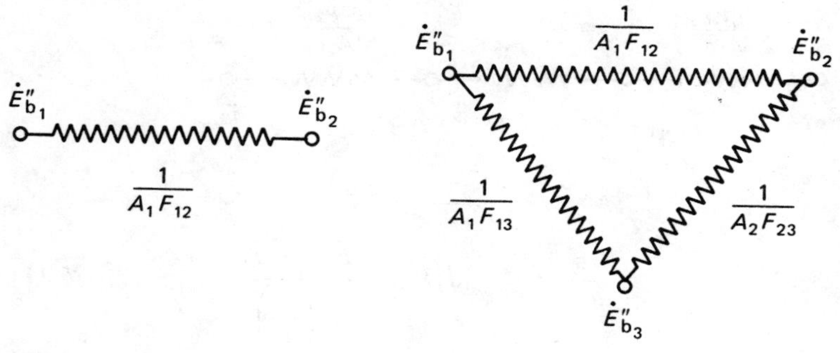

Figure 5.5

An analogy is made to Ohm's law ($I = V/R$) such that if

$$I \text{ is analogous to } \dot{Q}_{12}$$

and

$$V \text{ is analogous to } \sigma(T_1^4 - T_2^4) = (\dot{E}''_{b_1} - \dot{E}''_{b_2})$$

then

$$\frac{1}{A_1 F_{12}} \text{ must be analogous to resistance.}$$

Figure 5.5 shows the electrical circuit diagram with potentials $\dot{E}''_{b_1}$ and $\dot{E}''_{b_2}$ at either end of a single resistance. Figure 5.5 also shows a three black body problem and from the analogy the heat transfer rates can be directly written as

$$\dot{Q}_{12} = \frac{\dot{E}''_{b_1} - \dot{E}''_{b_2}}{\dfrac{1}{A_1 F_{12}}}; \qquad \dot{Q}_{13} = \frac{\dot{E}''_{b_1} - \dot{E}''_{b_3}}{\dfrac{1}{A_1 F_{13}}};$$

$$\dot{Q}_{23} = \frac{\dot{E}''_{b_2} - \dot{E}''_{b_3}}{\dfrac{1}{A_2 F_{23}}}$$

For grey body radiation the electrical analogy must also allow for emissivity by replacing $\dot{E}''_b$ with $\dot{J}''$ the radiosity. This is achieved by introducing another resistance to alter the potentials at each node. The net rate of radiation leaving a grey surface of area A_1 will be

$$A_1(\dot{J}''_1 - \dot{G}''_1)$$

From equation (5.10) this becomes

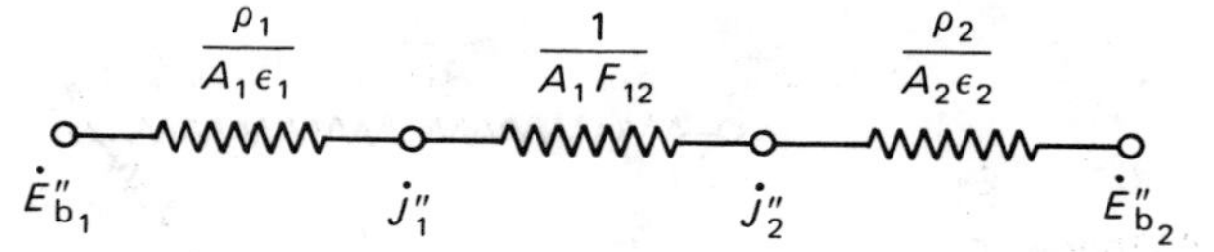

$\frac{\rho_1}{A_1\epsilon_1}$ $\frac{1}{A_1F_{12}}$ $\frac{\rho_2}{A_2\epsilon_2}$

$\dot{E}''_{b_1}$ $\dot{J}''_1$ $\dot{J}''_2$ $\dot{E}''_{b_2}$

Figure 5.6

$$\frac{A_1\varepsilon_1}{\rho_1}(\dot{E}'''_{b_1} - \dot{J}''_1) = \frac{\dot{E}'''_{b_1} - \dot{J}''_1}{\left(\dfrac{\rho_1}{A_1\varepsilon_1}\right)} \tag{5.11}$$

Equation (5.11) is incorporated into the electrical analogy as shown in Figure 5.6 and, clearly,

$$\dot{Q}_{12} = \frac{\dot{E}'''_{b_1} - \dot{E}'''_{b_2}}{\Sigma\text{resistance}} = \frac{\sigma(T_1^4 - T_2^4)}{\dfrac{\rho_1}{A_1\varepsilon_1} + \dfrac{1}{A_1F_{12}} + \dfrac{\rho_2}{A_2\varepsilon_2}} \tag{5.12}$$

5.6 General multisurface grey body radiation

Figure 5.7 shows a network for three grey surfaces. To solve such problems Kirchoff's electrical law for currents is applied to each $\dot{J}''$node giving three simultaneous equations of the form shown below for node 1:

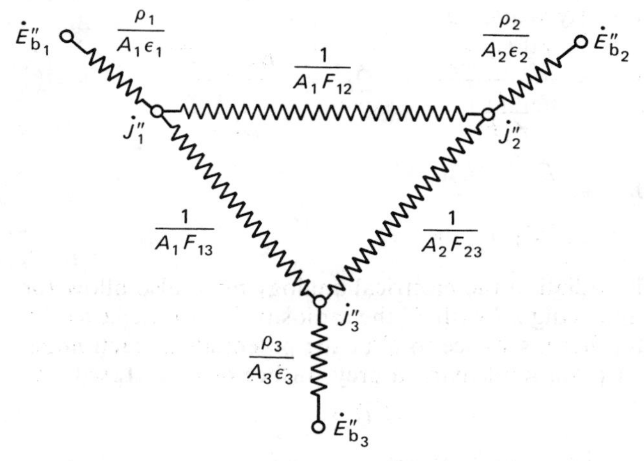

Figure 5.7

$$\frac{\dot{E}''_{b_1} - \dot{J}''_1}{\dfrac{\rho_1}{A_1 \varepsilon_1}} + \frac{\dot{J}''_2 - \dot{J}''_1}{\dfrac{1}{A_1 F_{12}}} + \frac{\dot{J}''_3 - \dot{J}''_1}{\dfrac{1}{A_1 F_{13}}} = 0 \qquad (5.13)$$

These three simultaneous equations are solved for $\dot{J}''_1$, $\dot{J}''_2$ and $\dot{J}''_3$ and heat transfer rates can be directly obtained. For example, the heat transfer rate from 1 to 2:

$$\dot{Q}_{12} = \frac{\dot{J}''_1 - \dot{J}''_2}{\dfrac{1}{A_1 F_{12}}}$$

Equation (5.13) can be rearranged to give

$$\dot{J}''_1 - (1 - \varepsilon_1)F_{12}\dot{J}''_2 - (1 - \varepsilon_1)F_{13}\dot{J}''_3 = \varepsilon_1 \dot{E}''_{b_1}$$

or, in general for the ith equation involving N surfaces j:

$$\dot{J}''_i - (1 - \varepsilon_i) \sum_{j=1}^{N} F_{ij}\dot{J}''_j = \varepsilon_i \dot{E}''_{b_i}$$

in which $F_{ii} = 0$ unless the surface can 'see' itself (concave). This will yield N equations which can be solved by Gaussian elimination or other appropriate method (Program 5.4).

5.7 Non-grey bodies

A body whose emissivity is not constant can be approximated by wavelength bands in which an average emissivity may be used. The heat transfer rate can then be calculated for each band and added to obtain the total heat transfer rate. The circuit resistances involving emissivity will be different for each band.

5.8 Special situations

1. If any surface is black then $\dot{J}'' = \dot{E}''_b$.
2. *Insulated* or *refractory* surfaces in a radiation network for which $\dot{Q} = 0$ do not exchange heat but influence the processes because they absorb and reradiate energy. The $\dot{J}''$ potential at such a node will represent the steady state temperature of the surface (Figure 5.8(a)), i.e. since $\dot{Q} = (\dot{E}''_b - \dot{J}'')/\rho/A\varepsilon = 0$ then $\dot{J}'' = \sigma T^4$.
3. Radiation shields are used to minimise radiation between surfaces. An example is shown in Figure 5.8(b).

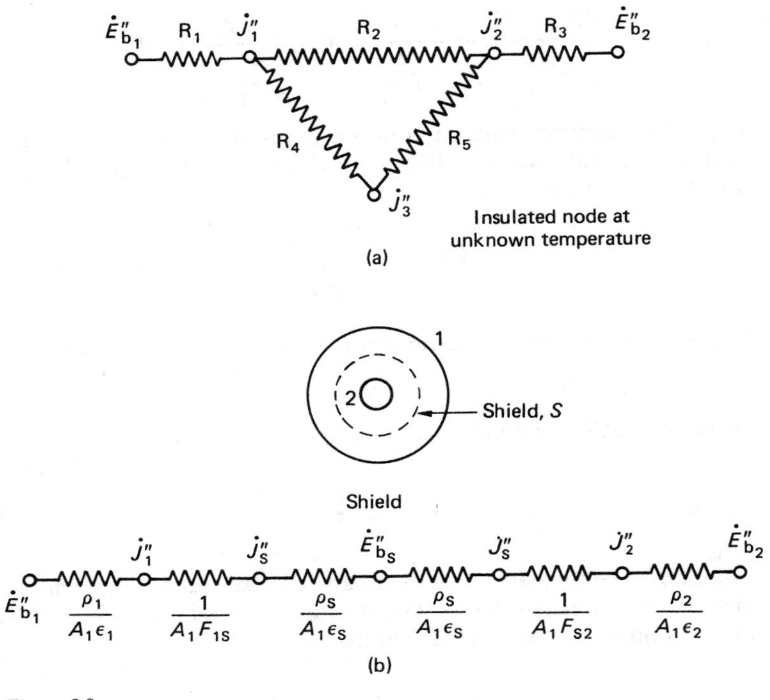

Figure 5.8

5.9 Gas radiation

In all discussions above it has been assumed that $\tau = 1$ for gaseous media. In many problems, particularly in combustion situations involving CO, CO_2, H_2O and SO_2, this is not true and further work is required [2].

References

1. Hamilton, D.C. and Morgan, W.R., *Radiant Interchange Configuration Factors*, NACA TN 2386 (1952).
2. White F.M., *Heat Transfer*, Addison-Wesley (1984).
3. Ozisik, M.N., *Heat Transfer*, McGraw-Hill (1985).

WORKED EXAMPLES

Program 5.1 Intensity of radiation

A fire at a temperature of 620 °C has a vertical plane area of 0.03 m^2.

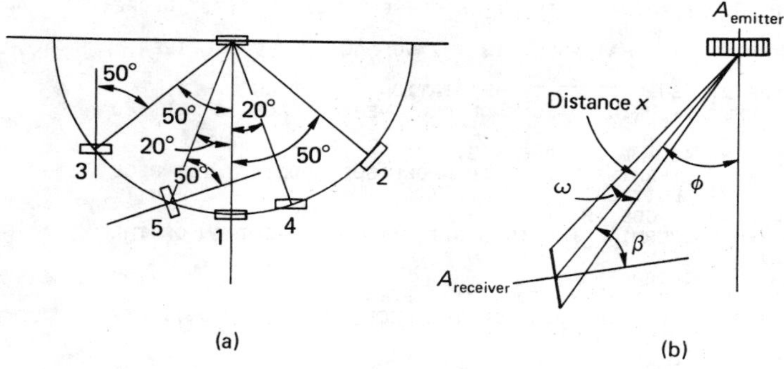

Figure 5.9

A person sitting by the fire is assumed to have a vertical plane area of 0.32 m² coaxial with the fire. Write a program to determine the energy incident on a person seated 1.5 m from the fire at the positions and inclinations 1–5 shown in Figure 5.9, these numbers 1–5 coinciding with the 'runs' of the program. The necessary reading for this program is Section 5.2.

From the definition of intensity of radiation and Lambert's law:

$$\mathrm{d}\dot{E}'' = i_\phi \mathrm{d}\omega = i_\mathrm{n} \cos\phi \cdot \mathrm{d}\omega$$

The receiver area is resolved to obtain the intercepting area to calculate the solid angle (Figure 5.9(b)):

$$\omega = \frac{A_{\mathrm{receiver}} \cos\beta}{x^2}$$

The emissive power $\dot{E}''$ of the emitter must be multiplied by the emitter area to determine the total radiation flux:

$$\Delta\dot{E} = \Delta\dot{E}'' \cdot A_{\mathrm{emitter}}$$

where $\Delta\dot{E}$ is the fraction of the total energy emitted per unit time that is intercepted by the receiver.

```
10  PRINT "DISTRIBUTION AND INTERCEPTION OF RADIATION"
20  REM  SET THE AREA OF THE RECEIVER, THE DISTANCE OF THE
 RECEIVER AND THE AREA AND TEMPERATURE OF THE EMITTER
30 AR = 0.32
40 X = 1.5
50 AE = 0.03
60 T = 893
70  REM  AREAS ARE IN M^2, DISTANCE IN M AND TEMPERATURE I
N K
80  PRINT " AT WHAT ANGLE TO THE NORMAL FROM THE EMITTER
"
85  PRINT "IS THE RECEIVER SITED?"
```

```
90  INPUT P: PRINT " DIRECTION OF RECEIVER IS ";P;" DEGRE
ES"
100  PRINT "AT WHAT ANGLE IS THE NORMAL TO THE RECEIVER "

105  PRINT "SURFACE TO THE RADIATION DIRECTION?"
110  INPUT B: PRINT "ANGLE OF RECEIVER SURFACE IS ";B;" D
EGREES"
120 B = 3.142 * B / 180:P = 3.142 * P / 180
130  REM  DETERMINE THE SOLID ANGLE SUBTENDED BY THE RECE
IVER AT THE EMITTER
140 O = AR *  COS (B) / X ^ 2
150  REM  DETERMINE THE INTENSITY IN THE DIRECTION OF THE
 RECEIVER
155 S = 0.0000000567
160 I = S * (T ^ 4) *  COS (P) / 3.142
170  REM  ENERGY INCIDENT ON THE RECEIVER IS PER UNIT ARE
A OF THE EMITTER
180 E = I * O
190 ET = E * AE
200 ET = ( INT (ET * 10 + 0.5)) / 10
210  PRINT " ENERGY INCIDENT ON RECEIVER IS ";ET;" W"
```

```
RUN
DISTRIBUTION AND INTERCEPTION OF RADIATION
 AT WHAT ANGLE TO THE NORMAL FROM THE EMITTER
IS THE RECEIVER SITED?
?0
 DIRECTION OF RECEIVER IS 0 DEGREES
AT WHAT ANGLE IS THE NORMAL TO THE RECEIVER
SURFACE TO THE RADIATION DIRECTION?
?0
ANGLE OF RECEIVER SURFACE IS 0 DEGREES
 ENERGY INCIDENT ON RECEIVER IS 49 W

]RUN
DISTRIBUTION AND INTERCEPTION OF RADIATION
 AT WHAT ANGLE TO THE NORMAL FROM THE EMITTER
IS THE RECEIVER SITED?
?50
 DIRECTION OF RECEIVER IS 50 DEGREES
AT WHAT ANGLE IS THE NORMAL TO THE RECEIVER
SURFACE TO THE RADIATION DIRECTION?
?0
ANGLE OF RECEIVER SURFACE IS 0 DEGREES
 ENERGY INCIDENT ON RECEIVER IS 31.5 W

]RUN
DISTRIBUTION AND INTERCEPTION OF RADIATION
 AT WHAT ANGLE TO THE NORMAL FROM THE EMITTER
IS THE RECEIVER SITED?
?50
 DIRECTION OF RECEIVER IS 50 DEGREES
AT WHAT ANGLE IS THE NORMAL TO THE RECEIVER
SURFACE TO THE RADIATION DIRECTION?
?50
ANGLE OF RECEIVER SURFACE IS 50 DEGREES
 ENERGY INCIDENT ON RECEIVER IS 20.2 W

]RUN
DISTRIBUTION AND INTERCEPTION OF RADIATION
 AT WHAT ANGLE TO THE NORMAL FROM THE EMITTER
IS THE RECEIVER SITED?
?20
 DIRECTION OF RECEIVER IS 20 DEGREES
AT WHAT ANGLE IS THE NORMAL TO THE RECEIVER
SURFACE TO THE RADIATION DIRECTION?
?20
```

```
ANGLE OF RECEIVER SURFACE IS 20 DEGREES
 ENERGY INCIDENT ON RECEIVER IS 43.2 W

]RUN
DISTRIBUTION AND INTERCEPTION OF RADIATION
 AT WHAT ANGLE TO THE NORMAL FROM THE EMITTER
IS THE RECEIVER SITED?
?20
 DIRECTION OF RECEIVER IS 20 DEGREES
AT WHAT ANGLE IS THE NORMAL TO THE RECEIVER
SURFACE TO THE RADIATION DIRECTION?
?50
ANGLE OF RECEIVER SURFACE IS 50 DEGREES
 ENERGY INCIDENT ON RECEIVER IS 29.6 W
```

Program nomenclature

P angular position of receiver relative to radiator normal
B angular position of receiver relative to radiation direction
O solid angle subtended by receiver at emitter
I intensity of radiation at angle P
E intercepted radiation per unit area of emitter
ET energy incident on receiver

Program notes

(1) The REM statements should indicate how this program is organised.
(2) The results of the five runs show that it is best to sit centrally in front of the fire, facing the fire (Run 1, 49 W) as would be expected, but with several persons using the fire each should face the fire rather than sit parallel to the fire; compare Run 2 (31.5 W) with Run 3 (20.2 W) and Run 4 (43.2 W) with Run 5 (29.6 W).
(3) Clearly, arrays of radiant heaters should be fitted so that the targets are placed at the most advantageous positions. This would require considerable calculation if uniform heating was required in a paint dryer (Problem 5.8).

Program 5.2 Geometric factor

Figure 5.3 shows a geometric factor chart for two equal sized coaxial rectangular plates. Data for this chart is obtained by integrating equation (5.7). Write a program to determine this geometric factor. The necessary reading for this program is Section 5.3. The relation is

$$F_{12} = \frac{2}{\pi A \cdot B}\left\{\ln\left[\frac{(1 + A^2)(1 + B^2)}{(1 + A^2 + B^2)}\right]^{1/2}\right.$$

$$+ B(1 + A^2)^{1/2}\cdot\tan^{-1}\left[\frac{B}{(1 + A^2)^{1/2}}\right]$$

$$+ A(1 + B^2)^{1/2}\cdot\tan^{-1}\left[\frac{A}{(1 + B^2)^{1/2}}\right]$$

$$\left. - A\cdot\tan^{-1} A - B\cdot\tan^{-1} B\right\}$$

where $A = X/D$ and $B = Y/D$ (see Figure 5.3). Write a program to determine this geometric factor. The necessary reading for this program is Section 5.3.

```
10  PRINT "GEOMETRIC FACTOR FOR TWO EQUAL SIZED "
15  PRINT "PARALLEL RECTANGULAR SURFACES"
20  REM  GEOMETRIC FACTOR F12 IS THE FRACTION OF ENERGY E
MITTED BY 1 THAT IS INTERCEPTED BY 2
25  PRINT
30  PRINT " WHAT IS THE DISTANCE BETWEEN THE PLATES?"
40  INPUT D: PRINT "DISTANCE BETWEEN PLATES IS ";D;" M"
50  PRINT "WHAT IS THE LENGTH OF A PLATE?"
60  INPUT X: PRINT " LENGTH OF A PLATE IS ";X;" M"
70  PRINT "WHAT IS THE WIDTH OF A PLATE?"
80  INPUT Y: PRINT "WIDTH OF A PLATE IS ";Y;" M"
90 A = X / D:B = Y / D
100 E = ((1 + (A ^ 2)) * (1 + (B ^ 2))) / (1 + (A ^ 2) +
(B ^ 2))
110 F = E ^ 0.5
120 G = (1 + (A ^ 2)) ^ 0.5
130 H = (1 + (B ^ 2)) ^ 0.5
140 I = B / G:J = A / H
150 L = 2 / (3.142 * A * B)
160 K = ( LOG (F) + B * G *  ATN (I) + A * H *  ATN (J) -
 A *  ATN (A) - B *  ATN (B))
170 GF = L * K
180  PRINT
190  PRINT " GEOMETRIC FACTOR F12 IS ";GF

]RUN
 GEOMETRIC FACTOR FOR TWO EQUAL SIZED
PARALLEL RECTANGULAR SURFACES

 WHAT IS THE DISTANCE BETWEEN THE PLATES?
?2.5
DISTANCE BETWEEN PLATES IS 2.5 M
WHAT IS THE LENGTH OF A PLATE?
?5
 LENGTH OF A PLATE IS 5 M
WHAT IS THE WIDTH OF A PLATE?
?5
WIDTH OF A PLATE IS 5 M

 GEOMETRIC FACTOR F12 IS .415199448
```

Program nomenclature

D distance between plates
X width of a plate
Y depth of a plate
GF geometric factor

Program notes

(1) The only problem in writing this program is the lengthy equation for F_{12}.
(2) Lines 100–140 determine variables E, F, G, H, I and J to enable the five terms contained within the bracket of equation (5.15) to be determined as variable K. Although it should be possible to program K directly, the use of simple steps enables arithmetic checks of each small piece.
(3) Simple statements make errors easier to detect.
(4) Several programs could be written to enable a library of geometric factors to be established (Problems 5.1 and 5.2).

Program 5.3 Radiation heat transfer between two grey surfaces

Two coaxial rectangular plates, each 5 m long by 5 m wide, are arranged parallel to each other 2.5 m apart. One plate is at 800 K with emissivity 0.8 and the other is at 300 K with emissivity 0.6. Write a program to determine the heat transfer rate between the plates.

This is a two grey body problem with three thermal resistances in series to which equation (5.12) and Figure 5.6 apply. Thus

$$\dot{Q}_{12} = \frac{\sigma(T_1^4 - T_2^4)}{\dfrac{\rho_1}{A_1\varepsilon_1} + \dfrac{1}{A_1F_{12}} + \dfrac{\rho_2}{A_2\varepsilon_2}}$$

and F_{12} is required. By choosing the same dimensions as those of Program 5.2 the geometric factor can be put in directly. The necessary reading for this program is Sections 5.4 and 5.5.

```
10  PRINT "RADIATION HEAT TRANSFER BETWEEN"
20  PRINT "TWO EQUAL PLANE PARALLEL SURFACES"
30  REM  INPUT THE EMISSIVITY,TEMPERATURE IN KELVIN AND A
REA IN M^2 FOR EACH PLANE
40 E1 = 0.8:E2 = 0.6
50 T1 = 800:T2 = 300
60 A = 25
70  REM INPUT THE GEOMETRIC FACTOR OBTAINED FROM THE PREV
IOUS PROGRAM
80 F = 0.415
```

```
90  REM  CALCULATE THE TOTAL RESISTANCE
100 R = ((1 - E1) / (A * E1)) + 1 / (A * F) + ((1 - E2) /
 (A * E2))
110  REM  CALCULATE THE EMISSIVE POWERS
120 S = 0.0000000567
130 P1 = S * (T1 ^ 4):P2 = S * (T2 ^ 4)
140 Q = (P1 - P2) / R
150 Q = ( INT (Q * 10 + 0.5)) / 10
160  PRINT "HEAT TRANSFER RATE IS ";Q;" W"
]RUN
RADIATION HEAT TRANSFER BETWEEN
TWO EQUAL PLANE PARALLEL SURFACES
HEAT TRANSFER RATE IS 171098.6 W
```

Program nomenclature

E1, E2	plate emissivities
T1, T2	plate temperatures
A	plate area
F	geometric factor
R	total thermal resistance
Q	heat transfer rate

Program notes

(1) The REM statements indicate how this program is organised.
(2) If the geometric factor were not arranged to be available line 80 would need alteration and the program would need to be expanded. How would this be done?
(3) The 171 kW heat transfer is spread over 25 m^2 of surface. On average this is 6.84 kW/m^2. Is the 171 kW spread uniformly? Problem 5.3 should be solved as an aid to this answer.

Program 5.4 General grey body program

Write a general grey body program to allow for a number of surfaces enclosed by black surroundings which will utilise the general equation (5.14). Figure 5.7 is relevant:

$$\dot{J}''_i - (1 - \varepsilon_i) \sum_{j=1}^{n} F_{ij}\dot{J}''_j = \varepsilon_i \dot{E}''_{b_i} \qquad (5.14)$$

This equation will produce a set of simultaneous equations for solution by Gaussian elimination (Program 2.2). The necessary reading for this program is Sections 5.3 and 5.6. The particular program is again set to use the geometric factor data from Program 5.2.

A room is heated by radiation from a warm ceiling. The room is 5 m square and 2.5 m high. Calculate the heat transfer rate from ceiling to floor (other heat transfer rates can also be determined from the solution).

Data

Surface 1; ceiling 50 °C, $\varepsilon = 0.8$, $A = 25\,m^2$
Surface 2; floor 20 °C, $\varepsilon = 0.7$, $A = 25\,m^2$
Surface 3; walls 18 °C, $\varepsilon = 1$, $A = 50\,m^2$ in total.

Thus from the earlier program 5.2 we obtain F_{12} as 0.41. Then from article 5.3 with

$$F_{12} = 0.41, \qquad F_{13} = 1\text{–}0.41 = 0.59$$

also,

$$F_{21} = 0.41, \qquad F_{23} = 1\text{–}0.41 = 0.59$$

but

$$A_1 F_{13} = A_3 F_{31}$$

thus

$$F_{31} = \tfrac{25}{50} \times 0.59 = 0.295 = F_{32}$$

and

$$F_{33} = 1 - F_{32} - F_{31} = 0.41$$

```
5  PRINT "GENERAL GREY BODY PROBLEM"
10  DIM A(30,31),F(30,30),E(30),T(30),X(30)
15  REM  DATA IS REQUIRED FOR EMISSIVITY,TEMPERATURE AND
GEOMETRIC FACTORS TO USE THE GAUSSIAN ELIMINATION,WHEN TH
IS IS SUPPLIED THE PROGRAM CALCULATES THE EQUATION CONSTA
NTS.
16  PRINT "HOW MANY  SURFACES ARE INVOLVED"
17  INPUT R: PRINT " THERE ARE "R" SURFACES"
19 S = 0.0000000567
20  FOR N = 1 TO R
25  READ E(N): READ T(N) :
30  NEXT N
35  FOR J = 1 TO R:I = R + 1
40 A(J,I) = E(J) * S * (T(J) ^ 4)
45  NEXT J
50  REM  READ IN THE GEOMETRIC FACTORS
55  FOR J = 1 TO R: FOR I = 1 TO R
60  READ F(J,I)
65  NEXT I: NEXT J
70  REM  DETERMINE THE COEFFICIENTS FOR GAUSSIAN ELIMINAT
ION
75  FOR J = 1 TO R: FOR I = 1 TO R
80 A(J,I) = 1
85  NEXT I: NEXT J
90  FOR J = 1 TO R: FOR I = 1 TO R
95  IF I = J THEN  GOTO 105
100 A(J,I) =  - (1 - E(J)) * F(J,I)
105  NEXT I: NEXT J
108  GOTO 170
```

```
110  REM  CALCULATE THE HEAT TRANSFER FROM CEILING TO FLO
OR
115  PRINT " WHAT IS THE CEILING AREA?"
120  INPUT AR: PRINT "CEILING AREA IS ";AR;" M^2"
125 Q = (X(1) - X(2)) / (1 / (AR * F(1,2)))
130 Q = ( INT (Q * 10 + 0.5)) / 10
140  PRINT "HEAT TRANSFER FROM CEILING TO FLOOR IS ";Q;"
W"
150  GOTO 1000
170  FOR J = 1 TO R
180  FOR I = J TO R
190  IF A(I,J) <  > 0 THEN  GOTO 230
200  NEXT I
210  PRINT "SOLUTION NOT POSSIBLE"
220  GOTO 1000
230  FOR K = 1 TO R + 1
240 X = A(J,K)
250 A(J,K) = A(I,K)
260 A(I,K) = X
270  NEXT K
280 Y = 1 / A(J,J)
290  FOR K = 1 TO R + 1
300 A(J,K) = Y * A(J,K)
310  NEXT K
320  FOR I = 1 TO R
330  IF I = J THEN  GOTO 380
340 Y =  - A(I,J)
350  FOR K = 1 TO R + 1
360 A(I,K) = A(I,K) + Y * A(J,K)
370  NEXT K
380  NEXT I
390  NEXT J
400  PRINT
410  FOR I = 1 TO R
420 X(I) = A(I,R + 1)
430  NEXT I
440  GOTO 115
1000  END
1100  DATA 0.8,323
1110  DATA  0.7,293
1120  DATA  1,291
1130  DATA  0,0.41,0.59
1140  DATA  0.41,0,0.59
1150  DATA  0.295,0.295,0.41
```

```
]RUN
GENERAL GREY BODY PROBLEM
HOW MANY  SURFACES ARE INVOLVED
?3
 THERE ARE 3 SURFACES

 WHAT IS THE CEILING AREA?
?25
CEILING AREA IS 25 M^2
HEAT TRANSFER FROM CEILING TO FLOOR IS 1454.5 W
```

Program nomenclature

E(J) emissivity
T(J) temperature
A(J, I) matrix coefficient in Gaussian elimination

F(J, I) geometric factors
AR ceiling area
X(N) output from Gaussian elimination which are the unknown J values
Q heat transfer

Program notes

(1) In line 10 the F(J, I) data space in the DIM statement is larger because for N surfaces there are N(N − 1)/2 independent F values.
(2) Enclosing surroundings can always see themselves so that there will always be a geometric factor associated with them (F_{33}). In many problems the surroundings are considered black when the term $(1 - \varepsilon_i)F_{ii}\dot{J}''_j$ is zero. Thus when $i = j = 3$ (the walls), the coefficient for $\dot{J}''_3$ is $[1 + (1 - \varepsilon_3)F_{33}]$. Since $\varepsilon_3 = 1$ the coefficient is 1. In this program all the diagonal $i = j$ coefficients are left at unity by lines 80 and 95. Thus the program only applies to black surroundings, but could be amended to allow for grey surroundings.
(3) The program structure is as follows:

Lines 20–30 read in data from lines 1100–1120.
Lines 35–45 determine Gaussian constants $\varepsilon\sigma T^4$.
Lines 50–65 read in geometric factors from lines 1130–1150.
Lines 70–105 determine coefficients in equations (5.14) for $\dot{J}''_1$, $\dot{J}''_2$, $\dot{J}''_3$ values.
Lines 170–430 are Gaussian elimination to find X(1), X(2), X(3) which are the unknown $\dot{J}''_1$, $\dot{J}''_2$, $\dot{J}''_3$ values.
Lines 115–430 calculate heat transfer.

(4) Go through the formulation of the equation in lines 70–105. This is how equation (5.14) is used and is important.
(5) Amend the program to find other heat transfers.
(6) It is possible to replace the reading of geometric factors with the program of Problem 5.2 to form a piece of software.
(7) It is also possible to consider this problem as a six-surface problem; amend the program accordingly. In this case there are no surfaces that can see themselves.

PROBLEMS

(5.1) Figure 5.10 shows three situations for which the geometric factor equations are given below. Write three programs for these equations but read Problem 5.2 before starting to avoid line numbering and variable complications.

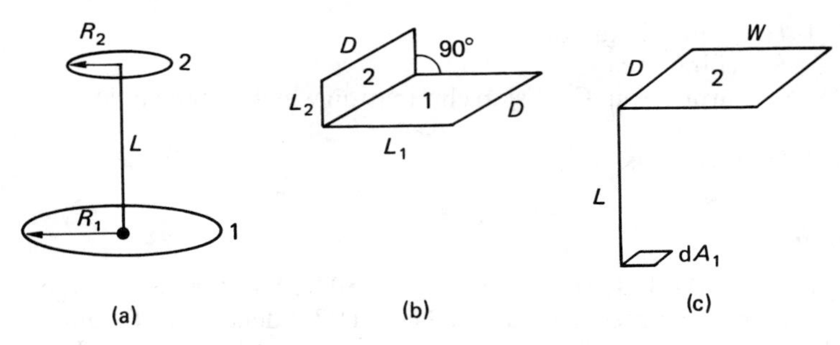

Figure 5.10

(a) For Figure 5.10(a), two discs:

$$F_{12} = \frac{1}{2}\left\{C - \left[C^2 - \frac{4B^2}{A^2}\right]^{1/2}\right\}$$

where

$$A = \frac{R_1}{L}, \quad B = \frac{R_2}{L} \text{ and } C = \left(\frac{1 + B^2}{A^2}\right) + 1$$

(b) For Figure 5.10(b), two planes at right-angles:

$$F_{12} = \frac{1}{\pi B}\Bigg(B \tan^{-1}\frac{1}{B} + A \tan^{-1}\frac{1}{A} - (A^2 + B^2)^{1/2} \tan^{-1}\frac{1}{(A^2 + B^2)^{1/2}}$$

$$+ \frac{1}{4}\ln\left\{\left[\frac{(1 + B^2)(1 + A^2)}{(1 + B^2 + A^2)}\right]\left[\frac{B^2(1 + B^2 + A^2)}{(1 + B^2)(A^2 + B^2)}\right]^{B^2} \times \left[\frac{A^2(1 + A^2 + B^2)}{(1 + A^2)(A^2 + B^2)}\right]^{A^2}\right\}\Bigg)$$

where

$$B = \frac{L_1}{D} \text{ and } A = \frac{L_2}{D}$$

(c) For Figure 5.10(c), a plane and a small area parallel to the plane:

$$F_{dA_1 2} = F_{12} = \frac{1}{2\pi}\left\{\frac{A}{(1 + A^2)^{1/2}} \tan^{-1}\left[\frac{B}{(1 + A^2)^{1/2}}\right] + \frac{B}{(1 + B^2)^{1/2}} \tan^{-1}\left[\frac{A}{(1 + B^2)^{1/2}}\right]\right\}$$

where

$$A = \frac{W}{L} \quad \text{and} \quad B = \frac{D}{L}$$

(5.2) Write a general geometric factor program incorporating the work of Program 5.2 and Problem 5.1 as four subroutines. Add any other programs available to form a piece of software. Line numbering should start at a convenient high value to avoid future complications, the number being chosen to fit in with any other subroutines which might be being used at the same time. An example might be fluid properties (Problem 4.1).

(5.3)

(a) A rectangular plate 5 m by 5 m is situated 2.5 m above a flat bar 5 m long and 0.1 m wide, the bar being placed symmetrically directly below the centre line of the plate (Figure 5.11). The plate is at a temperature of 800 K and the bar at 300 K. The emissivity of the plate is 0.8 and of the bar is 0.6. Using the geometric factor program written in Problems 5.1 or 5.2 for a small area and a plate and the addition rules of Figure 5.4, write a program to determine the distribution of heat transfer rate to the bar.

(b) A hall has a heated ceiling at a temperature of 50 °C. The ceiling height is 5 m and the hall is 12 m × 16 m. The floor temperature is 20 °C. The emissivity of the ceiling is 0.8 and of the floor is 0.7. Write a program to determine the heating distribution by radiation at floor level assuming that the walls of the hall may be considered black. Are the temperatures needed to determine this distribution? Is the distribution the same if the hall has a control system so that the ceiling temperature can be raised or lowered to any temperature between 30 °C and 60 °C?

(5.4) A rectangular, water cooled metal surface 2 m by 1.5 m is located 0.5 m directly above and parallel to a hot surface which is also 2 m by 1.5 m. The emissivity and temperature of the lower hot

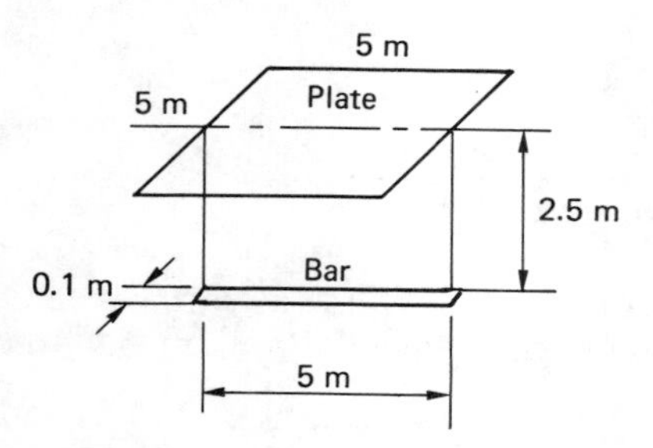

Figure 5.11

surface are 0.9 and 500 °C, respectively, and those of the upper cold surface are 0.7 and 20 °C. In order to reduce the heat transfer rate from the lower surface to the upper surface a radiation shield 2 m by 1.5 m is to be placed centrally between and parallel to the other surfaces. Write a program to determine the change in heat transfer rate and the temperature of the shield for various values of shield emissivity between 0.1 and 0.9. What happens if the shield is not placed centrally? (Section 5.8(iii) refers to radiation shields.)

(5.5) Two equal sized coaxial discs of diameter 2 m are placed 0.5 m apart in large surroundings. The discs exchange radiation with each other and with the surroundings through the gap between them. The temperatures and emissivities of the system components are

Disc 1	500 °C	0.8
Disc 2	100 °C	0.6
Surroundings	25 °C	large

Determine the heat transfer rate from the high temperature disc to the low temperature disc and the total heat transfer rate to the surroundings. There is no heat transfer from the 'back' of the discs.

(5.6)

(a) Program 5.5 may be amended to deal with non-black surroundings (see program note 2). Amend the program to allow the emissivity of the surroundings to be 0.9.

(b) Program 5.4 may be augmented to deal with surfaces of unknown temperature but at which the net heat transfer is known. The simplest case is an insulated surface for which $\dot{Q} = 0$ and Section 5.8(ii) showed how to obtain the surface temperature. For other cases when $\dot{Q}$ is not zero we may apply Kirchoff's law to the $\dot{J}_i''$ node to obtain

$$\dot{Q}_i = \frac{\dot{E}_{b_i}'' - \dot{J}_i''}{\dfrac{\rho_i}{A_i\varepsilon_i}} = \sum_{j=1}^{N}\left(\frac{\dot{J}_i'' - \dot{J}_j''}{\dfrac{1}{A_iF_{ij}}}\right)$$

then as $\Sigma\, \dot{J}_i'' F_{ij} = \dot{J}_i''$ we find

$$\dot{Q}_i = A_i\left(\dot{J}_i'' - \sum_{j=1}^{N} \dot{J}_j'' F_{ij}\right) \qquad (5.16)$$

Equation (5.16) is used instead of equation (5.14) at surfaces for which $\dot{Q}$ is known but T is unknown. Amend Program 5.4 to demonstrate this method by:

(i) changing lines 110–140 to determine the total heat transfer rate from the ceiling which must equal the heat input to the ceiling from the energy supply;

(ii) rearranging the program to use equation (5.16) for the ceiling node and hence determine the ceiling temperature.

(5.7) The plate in Program 5.3, which is at 300 K, is replaced by a new design made of plastic material. This material has an emissivity of 0.2 between wavelengths (λ) of 0 and 3 μm and of 0.8 for $\lambda > 3\ \mu$m. Write a program to determine the heat transfer rate between the plates. For solution it will be necessary to determine the fraction of the total black body radiation that lies in each wavelength band using Planck's law at a temperature of 300 K (Section 2.9).

Open-ended problems

(5.8) Many mass produced objects of various shapes (freezers, cycle frames, car bodies, etc.) are painted and a 'drying' process will be needed to evaporate the solvent and harden the paint. This would require a reasonably uniform distribution of heat over the surface without excessive surface temperature which might damage the finish. Radiation might be one way to achieve the objective. Drying could take place with the product in 'steady flow' through the booth or in batches of one or more for a fixed period of time within the booth. Some products need painting on several surfaces at different inclinations; some need all-over coating which may include internal surfaces. These are only some of the points which would need to be included in a decision about the positioning of radiating surfaces in a drying booth.

Consider the concept of paint drying by radiation heat transfer and draw up a list of the data required. When this is obtained write program(s) to determine the distribution and temperature of radiating surfaces to give uniform heating to bodies of various shapes.

(5.9) In Section 5.9 attention was drawn to the problems of gas radiation when transmissivity is not unity. In this case the gases will both emit and absorb radiation. Two important gases which fall in this group are water vapour and carbon dioxide which are both produced during the combustion of hydrocarbon fuel. These gases do not emit over a continuous spectrum of wavelengths but, rather, in narrow wavelength bands. The classical method of approach is by Hottel and Egbert [1] who produced charts for both water and carbon dioxide and Edwards [2] presents information on calculation techniques.

Make a study of the methods for radiation heat transfer in absorbing and emitting media to enable a program to be written to solve simple problems involving heat transfer from combustion products to furnace walls, tube banks, etc. [3].

References

1. Hottel, H.C. and Egbert, R.B., 'Radiant heat transmission from water vapor,' *Transactions of the American Institute of Chemical Engineers*, **38**, 531–68 (1942).
2. Edwards, D.K., *Radiation Heat Transfer Notes*, Hemisphere (1981).
3. Kern, D.W., *Process Heat Transfer*, McGraw-Hill (1984).

(**5.10**) Radiation from the sun is responsible for the energy supply to the earth. Make a study of solar radiation [1] and then write a program to design a solar collector to heat a home in various latitudes [2]. For this problem it will be necessary to study statistical solar data for the geographical area concerned as well as radiation concepts [3]. The home could be the one considered in Problem 2.3, sited in different locations. Because solar radiation is often interrupted for long periods by clouds and is not available during the night, energy storage techniques will also need consideration.

References

1. White, F.M., *Heat Transfer*, Addison-Wesley (1984).
2. Kreith, F. and Bohn, M.S., *Principles of Heat Transfer*, Harper (1986).
3. CIBS Guide, Sections A2 and A4.

Chapter 6

Finned surfaces

6.1 Introduction

Convection heat transfer is governed by the equation

$$\dot{Q} = hA\theta$$

With given values of h and θ the only way of increasing the heat transfer rate is to enlarge the surface area A. This may be achieved by extending the surface with projecting fins of various shapes. The most common types are strip fins on plate surfaces and annular fins on tubular surfaces; both types may have rectangular or tapered profiles. In many heat exchangers, particularly liquid to gas applications, the h values on the gas side are much smaller than those on the liquid side and fins would be added to the gas side until the thermal resistance of each are approximately equal. An example of this technique can be seen in the car radiator. The addition of fins alters flow patterns and h values and it is unlikely that h will be constant over the whole fin, but this effect is ignored in simple analysis. Account is, however, taken of the change of temperature along the fin length by a fin efficiency.

6.2 Strip fins of rectangular profile

Fins are considered long and thin enough for one-dimensional conduction laws to apply. Figure 6.1 shows a fin and by considering the balance between conduction and convection heat transfer to and from a small element distant x from the wall (root of the fin) a differential equation to the temperature distribution is established:

$$\frac{d^2\theta}{dx^2} - m^2\theta = 0$$

where $\theta = T - T_f$ is the temperature difference between fin and fluid at any point
$m = (ph/kA)^{1/2}$

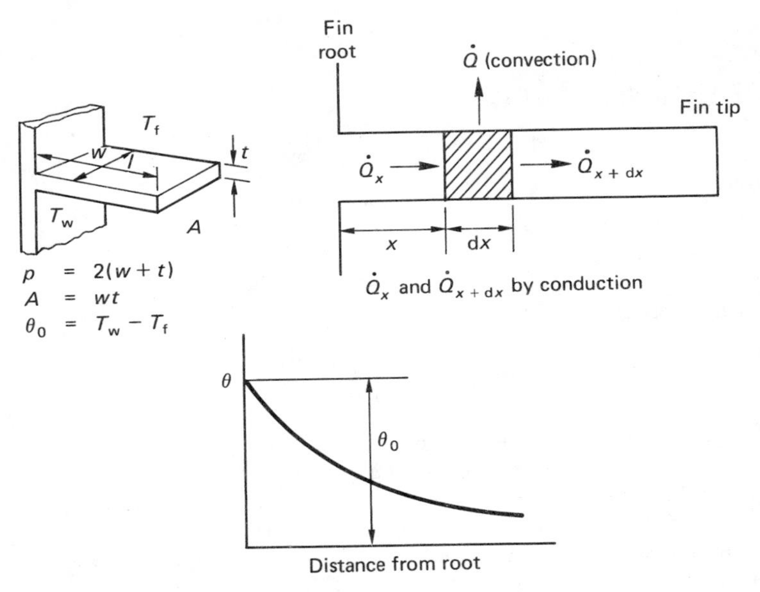

Figure 6.1

p is the fin perimeter
l is the fin length
k is the thermal conductivity of the fin material
A is the fin cross-sectional area
h is the average convection heat transfer coefficient
and T_w is the wall temperature.

The solution to this equation for long, thin fins with negligible heat loss from the tip is obtained by using the boundary conditions:

(a) at the wall where $x = 0$ and $\theta = \theta_0 = T_w - T_f$;

(b) at the tip where $x = 1$ and $\dfrac{d\theta}{dx} = 0$.

It is found that the temperature distribution is given by (Figure 6.1):

$$\frac{\theta}{\theta_0} = \frac{\cosh m(l - x)}{\cosh ml} \tag{6.1}$$

and hence the heat transfer rate from the fin is given by

$$\dot{Q} = mkA\theta_0 \tanh ml \tag{6.2}$$

Since this heat transfer is less than that which would occur if the whole fin was at the wall temperature a fin efficiency is defined by

$$\eta_{\text{fin}} = \frac{\text{actual heat transfer rate}}{\text{heat transfer rate if whole fin were at wall temperature}}$$

In this case

$$\eta_{\text{fin}} = \frac{\tanh ml}{ml} \tag{6.3}$$

This analysis also applies to pin fins with various cross-sections.

6.3 Annular fins and tapered fins

If a similar analysis is made for annular fins of rectangular profile (Figure 6.2) the resulting differential equation is

$$\frac{d^2\theta}{dr^2} + \frac{1}{r}\frac{d\theta}{dr} - \left(\frac{2h}{kb}\right)\theta = 0$$

Similarly for a strip fin (Figure 6.2) with tapered triangular profile, analysis yields

$$\frac{d^2\theta}{dx^2} + \frac{1}{x}\frac{d\theta}{dx} - \left(\frac{2hl}{kb}\right)\frac{\theta}{x} = 0$$

Both these equations can be solved using tables of Bessel functions [1] or by numerical integration by fourth-order Runge–Kutta scheme [2] or by finite difference techniques. However, because of the widespread use of fins the concept of fin efficiency is normally used and graphs of fin efficiency are plotted from solutions to the equations [3]. Figure 6.3 shows such a chart. The fin heat transfer

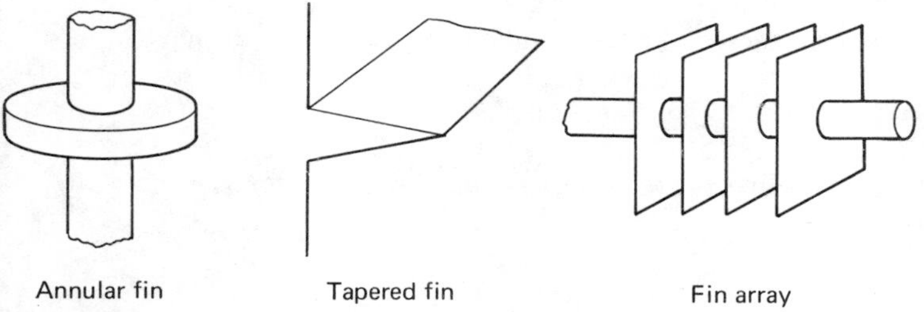

Figure 6.2

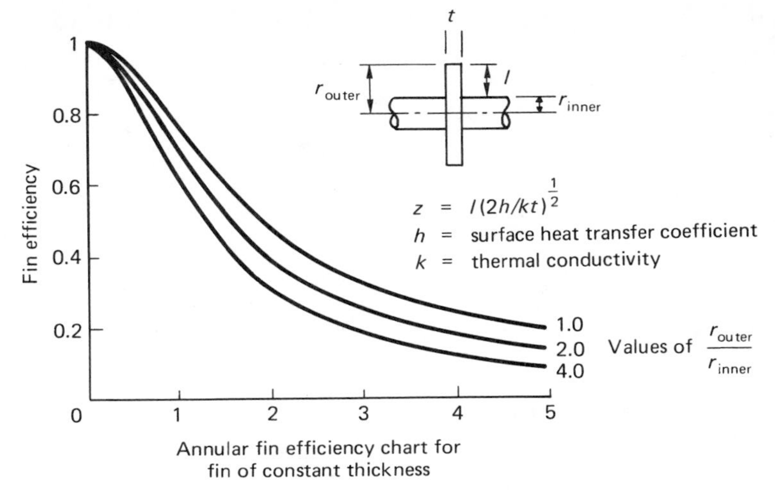

Figure 6.3

rate is then obtained from the definition of fin efficiency:

$$\eta_{fin} = \frac{\text{actual heat transfer rate}}{\text{heat transfer rate if whole fin were at wall temperature}}$$

6.4 Arrays of fins

Although the use of fins enhances the heat transfer area the extra area is inefficient due to the change in temperature along the surface which is expressed for the single fin by the fin efficiency. For arrays of fins (Figure 6.2) the effect can be calculated as shown below and for complex surfaces with fins (Figure 7.5) data is included on the chart for use in equations (6.5) and (6.6).

Let A_{fin} be the total fin area per unit length.

Let A be the total area *including* fin area per unit length:

$$\beta = \frac{A_{fin}}{A}$$

The total heat transfer rate from a finned surface is

$$\dot{Q}' = [\eta_{fin} A_{fin} + (A - A_{fin})]h\theta_0$$

where θ_0 is the temperature difference at the wall

$\dot{Q}'$ is the heat transfer rate per unit length
and h is the convection heat transfer coefficient

$$\dot{Q}' = \frac{[\eta_{\text{fin}} A_{\text{fin}} + (A - A_{\text{fin}})] h A \theta_0}{A}$$

$$\dot{Q}' = [\eta_{\text{fin}} \beta + (1 - \beta)] h A \theta_0 \quad (6.4)$$

Equation (6.4) may be written

$$\dot{Q}' = \eta' h A \theta_0 \quad \text{where} \quad \eta' = [\eta_{\text{fin}} \beta + (1 - \beta)] \quad (6.5)$$

In equation (6.5), η' is the *area weighted fin efficiency* which may be incorporated into the calculation of U value for heat exchangers. If U_A is the overall heat transfer coefficient *based on the enhanced surface area A* where the heat transfer coefficient is h_1, then

$$\frac{1}{U_A} = \frac{1}{\eta' h_1} + \frac{1}{\left(\frac{A_2}{A}\right) h_2} \quad (6.6)$$

where A_2 is the area of the unfinned side where the heat transfer coefficient is h_2 (Program 7.4).

6.5 Use of fins

Although fins increase the surface area the extra material increases the thermal resistance at the fin position and there may be situations where the heat transfer rate is not improved. As a guide, analysis shows that if $pk < hA$ fins will not enhance the heat transfer rate.

References

1. Eckert, E.R.G. and Drake, R.M. *Analysis of Heat and Mass Transfer*, McGraw-Hill (1972).
2. Adams, J.A. and Rogers, D.F. *Computer Aided Heat Transfer Analysis*, McGraw-Hill (1973).
3. Gardner, K.A. 'Efficiency of extended surfaces', *Transactions of the American Society of Mechanical Engineers*, **67**, 621–31 (1945).

WORKED EXAMPLES

Program 6.1 Heat transfer in a strip fin of rectangular profile

In the programs given in Chapter 3 the finite difference method was used to solve two-dimensional conduction problems using differen-

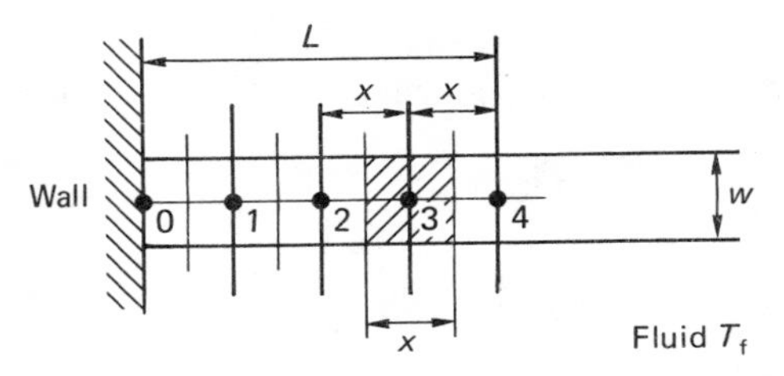

Figure 6.4

tial equations obtained by analysis. In the programs given in this chapter a similar method is used to solve one-dimensional conduction problems with convective boundaries. Clearly, there are alternative methods for this particular program because the fin we are considering can be solved by direct analysis and the resulting equations (6.1) and (6.2) could be programmed to give the temperature distribution in the fin and the heat transfer in the fin. However, this alternative is not always available and here we make a simple model of the fin in the computer and obtain an approximate solution without mathematical analysis.

Figure 6.4 shows a fin divided into four sections (five nodes). For this fin the wall temperature $T(0)$ is 100 °C and the surrounding fluid temperature is at 10 °C. The fin is 0.06 m long, 0.003 m thick and is made of material of thermal conductivity 50 W/m K. The surface heat transfer coefficient is 40 W/m² K. Consider unit width and write a program to determine the temperature distribution in the fin and the heat transfer rate from the fin.

Consider P nodes numbered from 0 to $(P - 1)$. At the wall where $P = 0$, $T(0) = 100$ and at the far end, where the node number $N = (P - 1)$, the conduction to the end must equal the convection from the side surfaces since in simple long fin theory there is no convection from the end. Thus

$$k(w \cdot 1)\left(\frac{T_N - T_{P-2}}{x}\right) + 2\left(\frac{x}{2} \cdot 1\right)h(T_f - T_N) = 0$$

$$T_N = \frac{\left(T_{P-2} + \left(\frac{hx^2}{kw}\right)T_f\right)}{\left(1 + \left(\frac{hx^2}{kw}\right)\right)}$$

The main body of the fin (nodes = 1 to $P - 2$) will be governed by the balance between conduction into the section, conduction out of the section and convection from the side surfaces:

$$k(w \cdot 1)\left(\frac{T_{N-1} - T_N}{x}\right) + k(w \cdot 1)\left(\frac{T_{N+1} - T_N}{x}\right)$$
$$+ 2h(x \cdot 1)(T_f - T_N) = 0$$

$$T_N = \frac{\left(T_{N-1} + T_{N+1} + 2\left(\frac{hx^2}{kw}\right)T_f\right)}{2\left(1 + \left(\frac{hx^2}{kw}\right)\right)}$$

The heat transfer rate from the fin must be equal to the heat transfer into the fin at the wall element:

$$\dot{Q} = k(w \cdot 1)\left(\frac{T_0 - T_1}{x}\right) + 2\left(\frac{x}{2} \cdot 1\right)h\left(\left(\frac{T_0 + T_1}{2}\right) - T_f\right)$$

$$\dot{Q} = k(w \cdot 1)\left(\frac{T_0 - T_1}{x}\right) + \left(\frac{hx}{2}\right)(T_0 + T_1 - 2T_f)$$

The necessary reading for this program is Section 6.2.

```
10  PRINT " LONGITUDINAL FIN"
20  PRINT
30  DIM T(10)
35 C = 10000000
40  PRINT "HOW MANY NODES DO YOU CHOOSE?"
50  INPUT P: PRINT "THERE ARE ";P;" NODES"
55  REM  GUESS THE INITIAL TEMPERATURES
58 T(0) = 100
60  FOR N = 1 TO (P - 1)
70 T(N) = (T(0) + TF) / 2
80  NEXT N
90  PRINT " WHAT IS THE FIN LENGTH?"
100  INPUT L: PRINT "FIN LENGTH IS";L;" M"
110  PRINT "WHAT IS THE FIN THICKNESS?"
120  INPUT W: PRINT "FIN THICKNESS IS";W;" M"
130  PRINT "WHAT IS THE THERMAL CONDUCTIVITY?"
140  INPUT K: PRINT "THERMAL CONDUCTIVITY IS";K;" W/M K"
150  PRINT "WHAT IS THE SURFACE HEAT TRANSFER COEFFICIENT
?"
160  INPUT H: PRINT "HEAT TRANSFER COEFFICIENT IS";H;" W/
M^2 K"
170  PRINT "WHAT IS THE FLUID TEMPERATURE?"
180  INPUT TF: PRINT "FLUID TEMPERATURE IS";TF;" C"
190 X = L / (P - 1)
200 F = (H * (X ^ 2)) / (K * W)
210  REM  ITERATION ROUTINE TO FIND TEMPERATURE DISTRIBUT
ION
220 A = 0
230  FOR N = 1 TO (P - 2)
240 T(N) = (T(N - 1) + T(N + 1) + 2 * F * TF) / (2 * (1 +
 F))
250  NEXT N
260 N = (P - 1)
270 T(N) = (T(P - 2) + F * TF) / (1 + F)
280 A = A + 1
290  PRINT "NUMBER OF ITERATIONS";A
```

```
310 Q = (K * W * 1 * (T(0) - T(1))) / X + H * (X / 2) * 1
 * (T(0) + T(1) - 2 * TF)
320  IF (C - Q) < 0.1 THEN  GOTO 335
325 C = Q
330  GOTO 230
335  PRINT
340  PRINT "NODE NUMBER   TEMPERATURE IN C"
350  PRINT
360  FOR N = 0 TO (P - 1)
370 T(N) = ( INT (T(N) * 100 + 0.5)) / 100
380  PRINT  TAB( 6)N, TAB( 17)T(N)
390  NEXT N
400  PRINT
410  PRINT "THE NUMBER OF ITERATIONS IS ";A
420  PRINT
430  PRINT "THE HEAT TRANSFER FROM THE FIN IS ";Q;" W/M"
```

```
RUN
 LONGITUDINAL FIN

HOW MANY NODES DO YOU CHOOSE?
?5
THERE ARE 5 NODES
 WHAT IS THE FIN LENGTH?
?.06
FIN LENGTH IS.06 M
WHAT IS THE FIN THICKNESS?
?.003
FIN THICKNESS IS3E-03 M
WHAT IS THE THERMAL CONDUCTIVITY?
?50
THERMAL CONDUCTIVITY IS50 W/M K
WHAT IS THE SURFACE HEAT TRANSFER COEFFICIENT?
?40
HEAT TRANSFER COEFFICIENT IS40 W/M^2 K
WHAT IS THE FLUID TEMPERATURE?
?10
FLUID TEMPERATURE IS10 C
NUMBER OF ITERATIONS1
NUMBER OF ITERATIONS2
NUMBER OF ITERATIONS3
NUMBER OF ITERATIONS4
NUMBER OF ITERATIONS5
NUMBER OF ITERATIONS6
NUMBER OF ITERATIONS7
NUMBER OF ITERATIONS8
NUMBER OF ITERATIONS9
NUMBER OF ITERATIONS10
NUMBER OF ITERATIONS11
NUMBER OF ITERATIONS12
NUMBER OF ITERATIONS13
NUMBER OF ITERATIONS14
NUMBER OF ITERATIONS15
NUMBER OF ITERATIONS16
NUMBER OF ITERATIONS17

NODE NUMBER   TEMPERATURE IN C

     0          100
     1          77.51
     2          63.13
     3          55.14
     4          52.59

THE NUMBER OF ITERATIONS IS 17

THE HEAT TRANSFER FROM THE FIN IS 272.186489 W/M
```

Program nomenclature

C parameter to end program when iteration completed
P number of nodes
T(N) temperature at a node
TF fluid temperature
L fin length
W fin thickness
K fin thermal conductivity
H surface heat transfer coefficient
X length of a section
A iteration counter
Q heat transfer rate per unit width

Program notes

(1) The analytical solution to this problem gives a heat transfer rate of 275.4 W per unit width.
(2) The program structure is as follows:

Line 30 allows for up to 10 nodes.
Line 35 is used to set a value for C far in excess of the heat transfer rate to allow the first iteration (line 320).
Lines 40–200 set up the problem with lines 60–80 estimating the initial temperatures.
Lines 210–330 are the iteration routine used until the change in heat transfer rate is considered negligible, C being reset by line 325 to the latest value.
Lines 340–430 display the results.

(3) The program can be run with different numbers of nodes and different levels of accuracy: (C–Q) of line 325.

Program 6.2 An array of fins of the type used in Program 6.1

Fins are often used in arrays for which the fin pitch is known. Fin pitch is the distance between centre lines of adjacent fins. The total heat transfer rate for a finned surface can then be determined by one of two methods:

(a) by adding the heat transfer rate from the fins to the heat transfer rate from the unfinned area;
(b) by using the analysis of Section 6.4 to determine the area weighted fin efficiency (η') and hence equation (6.5) to determine the heat transfer rate per unit length of finned surface (unit width fins)

$$\dot{Q}' = \eta' h A \theta_0$$

The necessary reading for the program is Sections 6.2 and 6.4.

This program uses method (a) and determines the heat transfer rate from a single fin determined in Program 6.1 which is multiplied by the number of fins. The heat transfer rate from the unfinned area is then added:

$$\dot{Q}' = y\dot{Q}_{\text{fin}} + (1 - yw)h\theta_0$$

where y is the number of fins per unit length of finned surface and w is the fin thickness. (The surface is of unit width.)

```
500  PRINT " FIN ARRAY OVER 1 M LENGTH"
510  PRINT "USING THE FIN DESIGN ABOVE"
520  PRINT
530  PRINT "WHAT IS THE FIN PITCH?"
540  INPUT Z: PRINT "FIN PITCH IS ";Z;" M"
550 Y = 1 / Z
560  PRINT "FINS PER METRE= ";Y
570 Q = ((Q * Y) + ((1 - Y * W) * 1 * H * (T(0) - TF)))
580 Q = ( INT (Q * 100 + 0.5) / 100)
590  PRINT "HEAT TRANSFER FROM FINNED SURFACE= ";Q;" W/M^
2"
```

```
RUN
 LONGITUDINAL FIN

HOW MANY NODES DO YOU CHOOSE?
?5
THERE ARE 5 NODES
 WHAT IS THE FIN LENGTH?
?.06
FIN LENGTH IS.06 M
WHAT IS THE FIN THICKNESS?
?.003
FIN THICKNESS IS3E-03 M
WHAT IS THE THERMAL CONDUCTIVITY?
?50
THERMAL CONDUCTIVITY IS50 W/M K
WHAT IS THE SURFACE HEAT TRANSFER COEFFICIENT?
?40
HEAT TRANSFER COEFFICIENT IS40 W/M^2 K
WHAT IS THE FLUID TEMPERATURE?
?10
FLUID TEMPERATURE IS10 C
NUMBER OF ITERATIONS1
NUMBER OF ITERATIONS2
NUMBER OF ITERATIONS3
NUMBER OF ITERATIONS4
NUMBER OF ITERATIONS5
NUMBER OF ITERATIONS6
NUMBER OF ITERATIONS7
NUMBER OF ITERATIONS8
NUMBER OF ITERATIONS9
NUMBER OF ITERATIONS10
NUMBER OF ITERATIONS11
NUMBER OF ITERATIONS12
NUMBER OF ITERATIONS13
NUMBER OF ITERATIONS14
NUMBER OF ITERATIONS15
```

```
NUMBER OF ITERATIONS16
NUMBER OF ITERATIONS17

NODE NUMBER   TEMPERATURE IN C

     0           100
     1           77.51
     2           63.13
     3           55.14
     4           52.59

THE NUMBER OF ITERATIONS IS 17

THE HEAT TRANSFER FROM THE FIN IS 272.186489 W/M
 FIN ARRAY OVER 1 M LENGTH
USING THE FIN DESIGN ABOVE

WHAT IS THE FIN PITCH?
?.01
FIN PITCH IS .01 M
FINS PER METRE= 100
HEAT TRANSFER FROM FINNED SURFACE= 29738.65 W/M^2
```

Program nomenclature

Z fin pitch
Y fins/m
Q heat transfer rate

Program notes

(1) The program is very simple but it requires to be added to Program 6.1 to produce answers. (Program 6.1 determines the heat transfer from a single fin.)
(2) The program could be rewritten to use the second method (area weighted fin efficiency) to be of more general use for any fin shape (Problem 6.2).
(3) Is this program valid if Y is not an integer?

Program 6.3 Circumferential fin

This program considers an annular fin on a circular pipe. In such a fin the heat transfer area increases with increase of radius and an analysis yields a differential equation which can be solved with Bessel functions or by numerical integration. In this program, as in Program 6.1, we model the circumferential fin and use the simple one-dimensional heat transfer equation (2.4):

$$\dot{Q}' = \frac{2\pi k \Delta T}{\ln\left(\frac{r_2}{r_1}\right)} \qquad (2.4)$$

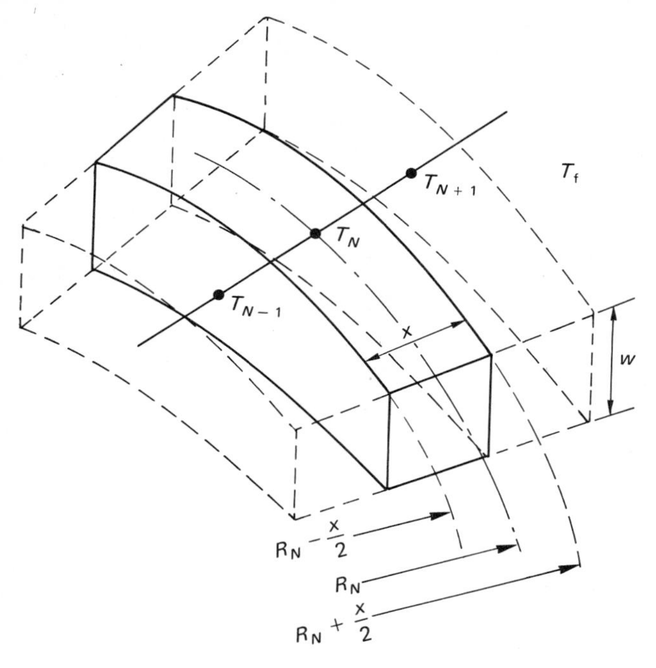

Figure 6.5

To use this we consider the fin as a series of annular rings (Figure 6.5) and by a balance between conduction into a ring, conduction out of a ring and convection from the ring surfaces we write

$$\frac{2\pi kw(T_{N-1} - T_N)}{\ln\left(\dfrac{R_N}{R_{N-1}}\right)} + \frac{2\pi kw(T_{N+1} - T_N)}{\ln\left(\dfrac{R_{N+1}}{R_N}\right)} + 2(2\pi R_N x)h(T_f - T_N) = 0$$

from which

$$T_N = \frac{T_{N-1}\ln\left(\dfrac{R_{N+1}}{R_N}\right) + T_{N+1}\ln\left(\dfrac{R_N}{R_{N-1}}\right) + 2R_N F\cdot\ln\left(\dfrac{R_{N+1}}{R_N}\right)\cdot\ln\left(\dfrac{R_N}{R_{N-1}}\right)T_f}{\ln\left(\dfrac{R_{N+1}}{R_N}\right) + \ln\left(\dfrac{R_N}{R_{N-1}}\right) + 2R_N F\cdot\ln\left(\dfrac{R_{N+1}}{R_N}\right)\cdot\ln\left(\dfrac{R_N}{R_{N-1}}\right)}$$

where $F = hx/kw$.

At the outside half-width ring by energy balance, ignoring convection from the outer ring edge surface, we obtain

$$\frac{2\pi kw(T_{N-1} - T_N)}{\ln\left(\dfrac{R_N}{R_{N-1}}\right)} + 2\left(2\pi\left(R_N - \frac{x}{4}\right)\frac{x}{2}h\right)(T_f - T_N) = 0$$

from which

$$T_N = \frac{T_{N-1} + F\left(R_N - \dfrac{x}{4}\right)\ln\left(\dfrac{R_N}{R_{N-1}}\right)T_f}{1 + F\left(R_N - \dfrac{x}{4}\right)\ln\left(\dfrac{R_N}{R_{N-1}}\right)}$$

The heat transfer rate at the wall is given by

$$\dot{Q} = \frac{2\pi kw(T_0 - T_1)}{\ln\left(\dfrac{R_1}{R_0}\right)} + 2\pi\left(R_0 + \frac{x}{4}\right)\frac{x}{2}h\left[\left(\frac{T_0 + T_1}{2}\right) - T_f\right]$$

The necessary reading for this program is in Sections 2.4 and 6.3.

Write a program to determine the heat transfer rate from an annular fin and test it with the following data:

Fin outer diameter	0.2 m
Fin root diameter	0.15 m
Fin thickness	2.5 mm
Thermal conductivity	48 W/m K
Surface heat transfer coefficient	280 W/m^2 K
Fin root temperature	170 °C
Ambient temperature	25 °C

```
10  PRINT "CIRCUMFERENTIAL FIN"
20  PRINT
30 C = 10000000
40  DIM T(20): DIM R(20)
50  PRINT "HOW MANY NODES DO YOU CHOOSE?"
60  INPUT P: PRINT "THERE ARE";P;" NODES"
70  REM  GUESS THE INITIAL TEMPERATURES
80  PRINT "WHAT IS THE WALL TEMPERATURE?"
85  INPUT T:T(0) = T: PRINT "WALL TEMPERATURE IS ";T(0);"
 C"
90  FOR N = 1 TO (P - 1)
100 T(N) = (T(0) + TF) / 2
110  NEXT N
120  PRINT "WHAT IS THE FIN ROOT DIAMETER?"
130  INPUT DR: PRINT "FIN ROOT DIAMETER IS ";DR;" M"
140  PRINT "WHAT IS THE FIN TIP DIAMETER?"
150  INPUT DT: PRINT "FIN TIP DIAMETER IS ";DT;" M"
160  PRINT "WHAT IS THE FIN THICKNESS?"
170  INPUT W: PRINT "FIN THICKNESS IS ";W;" M"
180  PRINT "WHAT IS THE THERMAL CONDUCTIVITY?"
190  INPUT K: PRINT "THERMAL CONDUCTIVITY IS ";K;" W/M K"
```

```
200  PRINT "WHAT IS THE SURFACE HEAT TRANSFER COEFFICIENT
?"
210  INPUT H: PRINT "HEAT TRANSFER COEFFICIENT IS ";H;" W
/M^2 K"
220  PRINT "WHAT IS THE FLUID TEMPERATURE?"
230  INPUT TF: PRINT "FLUID TEMPERATURE IS ";TF;" C"
235 R(0) = DR / 2:R(P - 1) = DT / 2
240 X = (DT - DR) / (2 * (P - 1))
250 F = (H * (X ^ 2)) / (K * W)
255 F = F / X
260  REM  ITERATION ROUTINE TO FIND TEMPERATURE DISTRIBUT
ION
270 A = 0
280  FOR N = 1 TO (P - 2)
290 R(N) = (DR / 2) + (N * X)
300  NEXT N
310  FOR N = 1 TO (P - 2)
320 LL = R(N + 1) / R(N)
330 MM = R(N) / R(N - 1)
340 E = (2 * R(N) * F *  LOG (LL) *  LOG (MM))
350 T(N) = (T(N - 1) *  LOG (LL) + T(N + 1) *  LOG (MM) +
 E * TF) / ( LOG (LL) +  LOG (MM) + E)
360  NEXT N
370 N = (P - 1)
380 I =  LOG (MM) * F * (R(N) - (X / 4))
390 T(N) = (T(N - 1) + I * TF) / (1 + I)
400 A = A + 1
410  PRINT "NUMBER OF ITERATIONS ";A
420 Q = ((2 * 3.142 * K * W * (T(0) - T(1))) / ( LOG (R(1
) / R(0)))) + H * 3.142 * (DR + (X / 2)) * (X / 2) * (T(0
) + T(1) - 2 * TF)
430  IF (C - Q) < 0.1 THEN  GOTO 460
440 C = Q
450  GOTO 310
460  PRINT
470  PRINT "NODE NUMBER    RADIUS IN M   TEMPERATURE IN C
 "
480  PRINT
490  FOR N = 0 TO (P - 1)
500 R(N) = ( INT (R(N) * 1000 + 0.5)) / 1000:T(N) = ( INT
 (T(N) * 100 + 0.5)) / 100
510  PRINT  TAB( 6)N, TAB( 17)R(N), TAB( 28)T(N)
520  NEXT N
530  PRINT "THE NUMBER OF ITERATIONS IS ";A
540  PRINT
550  PRINT " THE HEAT TRANSFER RATE FROM THE FIN IS ";Q;"
 W"
560 BI = (H * (DT - DR)) / (2 * K)
570  PRINT "THE BIOT NUMBER IS ";BI

RUN
CIRCUMFERENTIAL FIN

HOW MANY NODES DO YOU CHOOSE?
?10
THERE ARE10 NODES
WHAT IS THE WALL TEMPERATURE?
?170
WALL TEMPERATURE IS 170 C
WHAT IS THE FIN ROOT DIAMETER?
?.15
FIN ROOT DIAMETER IS .15 M
WHAT IS THE FIN TIP DIAMETER?
?.20
FIN TIP DIAMETER IS .2 M
WHAT IS THE FIN THICKNESS?
?.0025
```

```
FIN THICKNESS IS 2.5E-03 M
WHAT IS THE THERMAL CONDUCTIVITY?
?48
THERMAL CONDUCTIVITY IS 48 W/M K
WHAT IS THE SURFACE HEAT TRANSFER COEFFICIENT?
?280
HEAT TRANSFER COEFFICIENT IS 280 W/M^2 K
WHAT IS THE FLUID TEMPERATURE?
?25
FLUID TEMPERATURE IS 25 C
NUMBER OF ITERATIONS 1
NUMBER OF ITERATIONS 2
NUMBER OF ITERATIONS 3
NUMBER OF ITERATIONS 4
NUMBER OF ITERATIONS 5
NUMBER OF ITERATIONS 6
NUMBER OF ITERATIONS 7
NUMBER OF ITERATIONS 8
NUMBER OF ITERATIONS 9
NUMBER OF ITERATIONS 10
NUMBER OF ITERATIONS 11
NUMBER OF ITERATIONS 12
NUMBER OF ITERATIONS 13
NUMBER OF ITERATIONS 14
NUMBER OF ITERATIONS 15
NUMBER OF ITERATIONS 16
NUMBER OF ITERATIONS 17
NUMBER OF ITERATIONS 18

NODE NUMBER    RADIUS IN M    TEMPERATURE IN C

     0           .075            170
     1           .078            145.13
     2           .081            125.38
     3           .083            109.83
     4           .086            97.76
     5           .089            88.59
     6           .092            81.89
     7           .094            77.33
     8           .097            74.66
     9           .1              73.75
THE NUMBER OF ITERATIONS IS 18

 THE HEAT TRANSFER RATE FROM THE FIN IS 564.748034 W
THE BIOT NUMBER IS .145833333
```

Program nomenclature

C	parameter to end program when iteration completed
P	number of nodes
T(N)	temperature at a node
R(N)	radius of a node
DR	fin root diameter
DT	fin tip diameter
W	fin thickness
K	fin thermal conductivity
H	surface heat transfer coefficient
TF	fluid temperature
X	width of radial slice

LL, MM, E, I convenient program variables
A iteration counter
Q heat transfer rate

Program notes

(1) The fin efficiency chart solution to this problem is 558 W. However checking the program against other problems suggests that the accuracy of the program may be influenced by the value of the Biot number which is evaluated in lines 560 and 570. Bearing in mind the work of Sections 2.13 and 3.8 this would not be surprising and would be worth further investigation.
(2) The program structure is as follows:

Line 30 sets a value of C far in excess of the heat transfer rate to allow the second iteration (line 430).
Line 40 allows for up to 20 nodes.
Lines 50–255 set up the problem with lines 90–110 estimating the initial temperatures.
Lines 260–450 contain the iteration routine, lines 320, 330, 340 and 350 being used to obtain simple variables. The routine continues till line 430 decides the problem is solved.
Lines 470–550 display the results.

(3) The number of nodes can be varied and different values of (C–Q) tested.
(4) The program illustrates how situations may often be modelled in a simple way but obviously clear thought is needed and assumptions made must be valid. There are debatable points in the solution, for example no convection from the outer edge surface is considered. Is it worth allowing for the small increment in radius $x/4$ in the outside equation? Such items can be examined by the user.
(5) If *very narrow* strips were chosen could the program be considered with the longitudinal fin program? This would avoid the need for log ratios by using

$$\dot{Q} = kA\frac{\Delta T}{\Delta x}$$

PROBLEMS

(6.1) Determine the effect of fin dimensions, number of nodes and iteration cut off values on the accuracy of Program 6.1. The influence of length to thickness ratio is important to establish since the boundary condition at the end of the fin may need to be modified to allow for end convection when a short, fat fin is considered. For this case analysis shows

$$\frac{\theta}{\theta_0} = \frac{\cosh m(l - x) + \left(\frac{h}{km}\right) \sinh m(l - x)}{\cosh ml + \left(\frac{h}{km}\right) \sinh ml}$$

and

$$\dot{Q} = mkA\theta_0 \left[\frac{\tanh ml + \left(\frac{h}{km}\right)}{1 + \left(\frac{h}{km}\right) \tanh ml} \right]$$

It would also be interesting to consider two-dimensional effects in a short, fat fin using the methods of Chapter 3.

(6.2) Write a program for a longitudinal fin array based on area weighted fin efficiency. Compare the results with those of Program 6.2.

(6.3) Amend Program 6.3 to deal with an array of circumferential fins.

(6.4) Consider the modelling of tapered strip fins. Figure 6.6 suggests one way of dealing with the varying cross-sectional area.

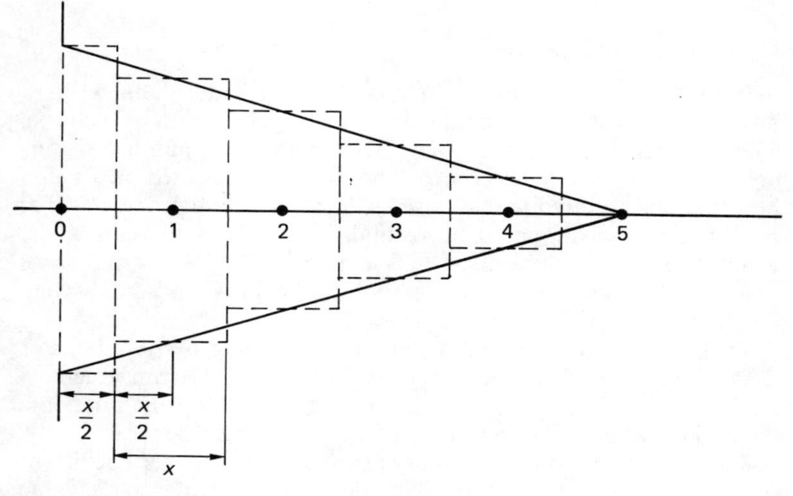

Figure 6.6

Compare the results with those obtained by using fin efficiency graphs and by programming the analytical equation of Section 6.3.
(6.5) Model a tapered circumferential fin and compare the results obtained with those obtained using fin efficiency graphs.
(6.6) Write a program for an array of tapered circumferential fins. Compare the heat transfer rates obtained with this program with those of Problem 6.3. Is there any advantage in the shape of the fin? Is one lighter than the other to give the same heat transfer? Is there an optimum fin shape to achieve a given heat transfer?
(6.7) A metal cylinder head on a small experimental engine is required to dissipate 10 kW by means of rectangular profile fins across the total width of the head which is 200 mm square. The fins may be up to 60 mm long. Write a program to determine a range of choices of length, thickness and pitch. The pitch should not be less than the fin thickness (see also Problem 6.10).

Data

Thermal conductivity of fin material	180 W/m K
Surface heat transfer coefficient	100 W/m^2 K
Surface temperature at fin root	230 °C
Ambient temperature	28 °C

(6.8) To achieve a given heat transfer rate with an array of circumferential fins with parallel faces is it better to use large or small outer diameter fins if the total weight of the array is to be minimised?

Open-ended problems

(6.9) A thin steel tube of approximately d mm outer diameter is to be manufactured with tapered strip fins along the length of the tube. The fin pitch is to be such that there is an exact number of fins around the outer circumference. The tube is to be used as a room heater and is required to dissipate 1 kW from each metre run by free convection when mounted horizontally as a 'skirting' heater. The water in the tube is at a mean bulk temperature of 70 °C and flows at a rate of 0.2 kg/s. The room temperature is 19 °C and the thermal conductivity of steel is 48 W/m K.

Write a program to find if a suitable design with *overall* diameter not exceeding 100 mm can be achieved. What other arrangement of fins on a tube could be considered for skirting heating? Evaluate the most likely alternative to find what size this would be.
(6.10) It has been assumed that the surface heat transfer coefficient over a fin surface is constant. The determination of such a value would require experiment in many cases. For example, annular fins

around a tube in cross-flow would experience complex flow and circumferential conduction would occur. The pitch of fins will also be important because flow is basically over a flat surface and the boundary layers developing on adjacent fins may interfere and cause the surface heat transfer coefficient to decrease.

Consider rectangular longitudinal fins and write a program to relate pitch, width, efficiency and heat transfer rate to boundary layer thickness. Assume that such a fin can be treated as a plate and that the hydrodynamic and thermal boundary layers both start at the leading edge of the fin.

For laminar flow over a flat plate with constant wall temperature:

$$\mathrm{Nu}_x = 0.332\,\mathrm{Re}_x^{1/2}\mathrm{Pr}^{1/3} \quad \text{for} \quad \mathrm{Re}_x < 5 \times 10^5, \quad \mathrm{Pr} > 0.1$$

and for turbulent flow:

$$\mathrm{Nu}_x = 0.0288\,\mathrm{Re}_x^{0.8}\mathrm{Pr}^{0.33} \quad \text{for} \quad \mathrm{Re}_x > 5 \times 10^5, \quad \mathrm{Pr} > 0.5$$

where Nu_x is the local Nusselt number hx/k and Re_x is $\rho Vx/\mu$. All properties are determined at the film temperature. The thickness of the hydrodynamic boundary layer is

$$\delta_x = \frac{4.96x}{\mathrm{Re}_x}$$

and of the thermal boundary layer is

$$\delta_t = \frac{\delta_x}{\mathrm{Pr}^{1/3}}$$

where x is the distance from the leading edge.

If a set of fins is to be fitted to the 'sides' of a motor cycle engine exposed to air flow, determine a suitable fin size and pitch to deal with the worst expected running conditions.

Chapter 7

Heat exchangers

7.1 Introduction

The most common form of heat exchanger (recuperator) consists of a solid, good conducting boundary separating a hot fluid from a cold fluid. The solid boundary has negligible thermal resistance so that the overall heat transfer coefficient between the fluids is given by

$$\frac{1}{U} = \sum \frac{1}{h} \quad \text{or} \quad \frac{1}{U'} = \sum \frac{1}{2\pi r h} \tag{7.1}$$

These heat exchangers (Figure 7.1) may be broadly classified by their construction (tubular including shell and tube, plate, complex or extended surface, etc.) and their flow pattern (single pass, multiple pass, parallel flow, counterflow, cross-flow). There is also a less common type of heat exchanger (regenerator) discussed in Problem 7.11.

The analysis of heat exchangers is normally made by one of two approaches, the LMTD (log mean temperature difference) method or the NTU method.

7.2 The LMTD mehod

This classical approach is based on the use of equations (2.10) and (2.12)

$$\dot{Q} = UA\theta \quad \text{or} \quad \dot{Q} = U'L\theta$$

Figure 7.2 shows parallel and counterflow temperature variation in a heat exchanger. Since θ varies it is necessary to analyse this variation to obtain a suitable mean value. It is found that

$$\theta_{\text{LMTD}} = \frac{\theta_1 - \theta_2}{\ln\left(\dfrac{\theta_1}{\theta_2}\right)} \tag{7.2}$$

Where θ_1 and θ_2 are the temperature differences at either end and θ_{LMTD} is the log mean temperature difference. It is also found that with the same inlet and exit temperatures counterflow designs give

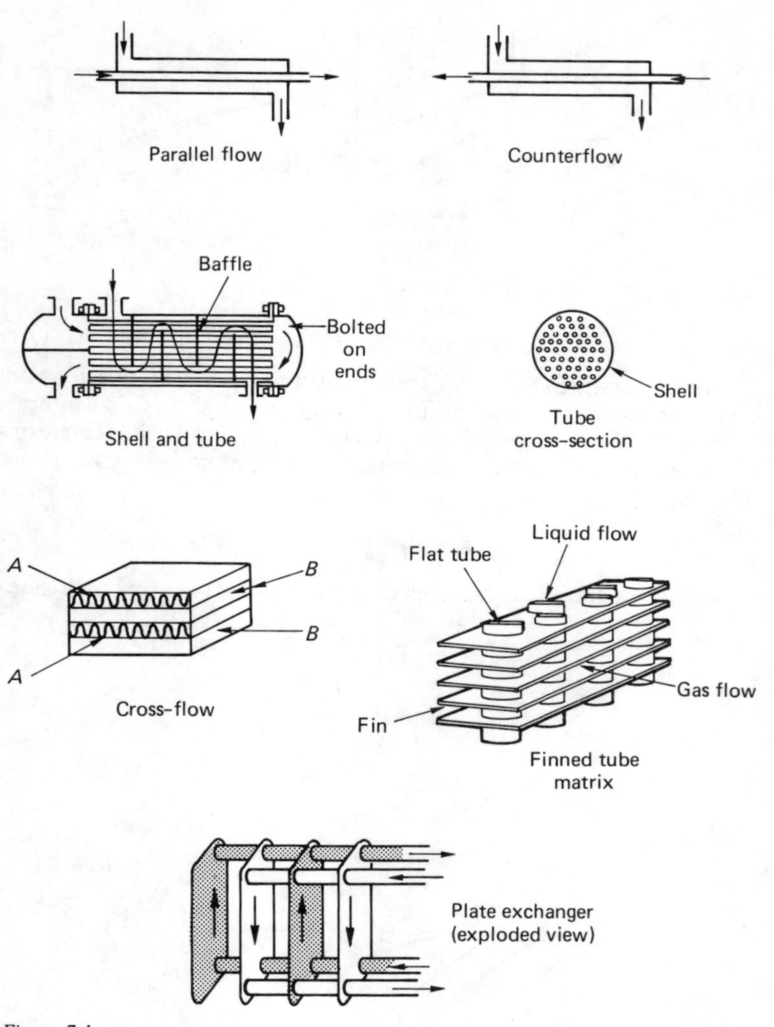

Figure 7.1

higher θ_{LMTD} than parallel flow designs and therefore use less area for heat transfer. Thus we write

$$\dot{Q} = UA\theta_{LMTD} \quad \text{or} \quad \dot{Q} = U'L\theta_{LMTD}$$

Heat exchangers that are neither counter nor parallel flow have been analysed [1] to give a correction factor F to the *counterflow* value of θ_{LMTD} so that, in general,

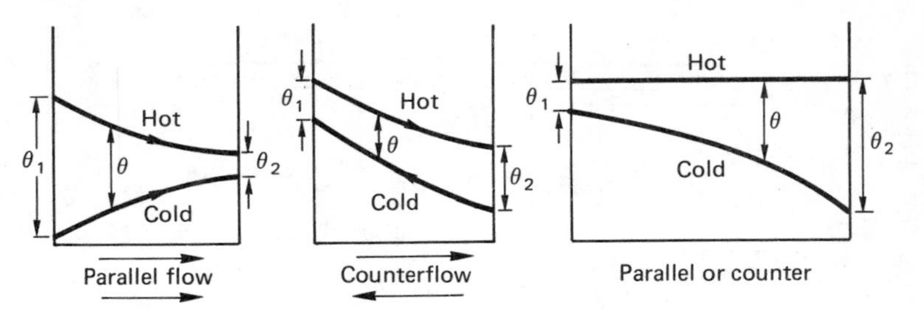

Figure 7.2

$$\dot{Q} = UAF\theta_{\text{LMTD}} \quad \text{or} \quad \dot{Q} = U'LF\theta_{\text{LMTD}} \tag{7.3}$$

Values of F (which will not exceed unity) for use in calculations are available as equations or graphs [2]. Figure 7.3 shows an example.

In order to solve a problem by the LMTD method using equations (7.1), (7.2) and (7.3) there are usually two other necessary relations:

(a) the energy balance

$$\dot{Q} = (\dot{m}c_p\Delta T)_{\text{hot}} = (\dot{m}c_p\Delta T)_{\text{cold}};$$

(b) the surface heat transfer coefficient equations (on each side)

$$\text{Nu} = \phi(\text{Re, Pr}) \quad \text{or} \quad \text{St} = \phi\left(\frac{f}{2}\right).$$

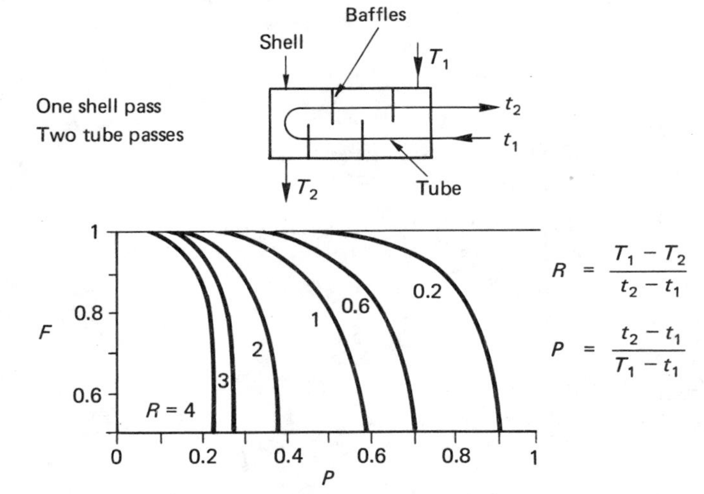

Figure 7.3

Then, *provided the inlet and outlet temperatures for each fluid are known and a value of U can be estimated*, the required area can be found and a possible design chosen. This choice enables the actual U value to be determined and iteration can proceed until convergence occurs (Program 7.2). Usually only inlet temperatures are known and an outlet temperature also needs to be guessed requiring further complicating iteration. (Student short examples do not suffer from such real life problems!)

7.3 The NTU method

The iteration problem caused by ignorance of exit temperatures was alleviated by the introduction of the NTU (number of transient units) method, however iteration for estimated U and A values is still necessary. Three parameters are defined:

$$\text{Number of transfer units, NTU} = \frac{UA}{(\dot{m}c_p)_{min}}$$

$$\text{Capacity ratio, } C = \frac{(\dot{m}c_p)_{min}}{(\dot{m}c_p)_{max}}$$

$$\text{Effectiveness, } E = \frac{\dot{Q}}{\dot{Q}_{max}}$$

$$= \frac{\text{actual heat transfer rate}}{\text{maximum possible heat transfer rate}}$$

The maximum possible heat transfer rate would be achieved (with an infinite area heat exchanger) when the hot stream outlet temperature reaches the cold stream inlet temperature or when the cold stream outlet temperature reaches the hot stream inlet temperature (Figure 7.2):

$$\dot{Q}_{max} = (\dot{m}c_p)_{min}(T_{hot_{in}} - T_{cold_{in}})$$

The actual heat transfer rate is given by

$$\dot{Q} = (\dot{m}c_p\Delta T)_{hot} = (\dot{m}c_p\Delta T)_{cold} \qquad (7.4)$$

Thus

$$E = \frac{(\dot{m}c_p\Delta T)_{hot}}{(\dot{m}c_p)_{min}(T_{hot_{in}} - T_{cold_{in}})} = \frac{(\dot{m}c_p\Delta T)_{cold}}{(\dot{m}c_p)_{min}(T_{hot_{in}} - T_{cold_{in}})} \qquad (7.5)$$

Kays and London [3] state

$$E = \phi(NTU, C, \text{flow arrangement})$$

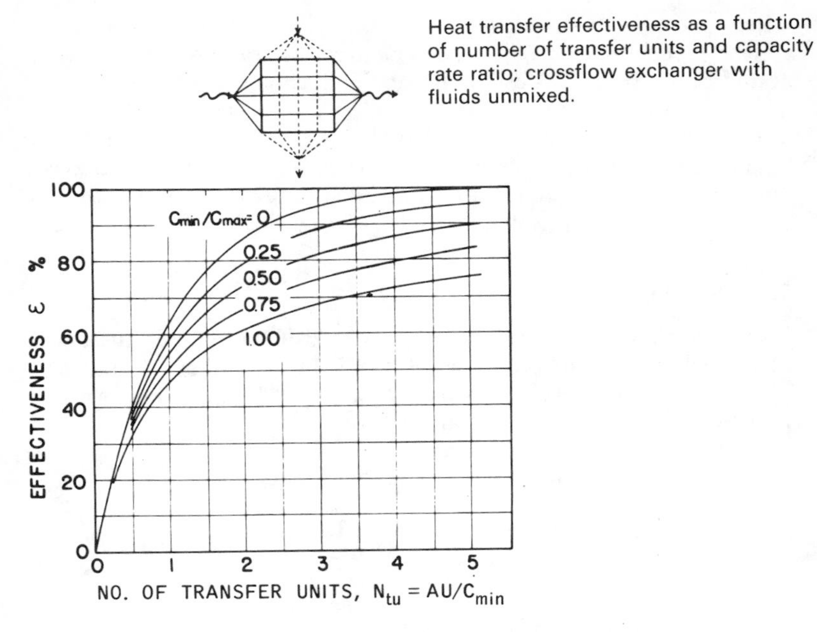

Figure 7.4 Reproduced with permission from Kays, W.M. and London, A.L., *Compact Heat Exchangers*, McGraw-Hill (1964)

and then give analyses for a range of heat exchangers to produce equations, tables and graphs relating *E*, *C* and *NTU*. Figure 7.4 shows a cross-flow example. To use the NTU method, known or estimated values of *U* and *A* enable *NTU* to be determined and *C* is also calculated. The graph or relation for the chosen configuration enables *E* to be found. Using *E* the outlet temperature and hence the heat transfer rate are calculated from equations (7.5) and (7.4). Iteration will be required for *U* and *A* to obtain satisfactory solution if estimates have been used. Clearly *E*, the effectiveness, should be high for a good design (Program 7.3).

7.4 Complex or extended surface heat exchangers

Heat exchangers such as a car radiator will not yield an algebraic solution to find surface heat transfer coefficients. Instead, standard forms are tested and the results presented on graphs or in tables. Kays and London [3] include data for a number of examples for use with the NTU method. Friction factors for pressure loss are also given in the data. Figure 7.5 shows a typical plot together with details of the matrix geometry (Program 7.4).

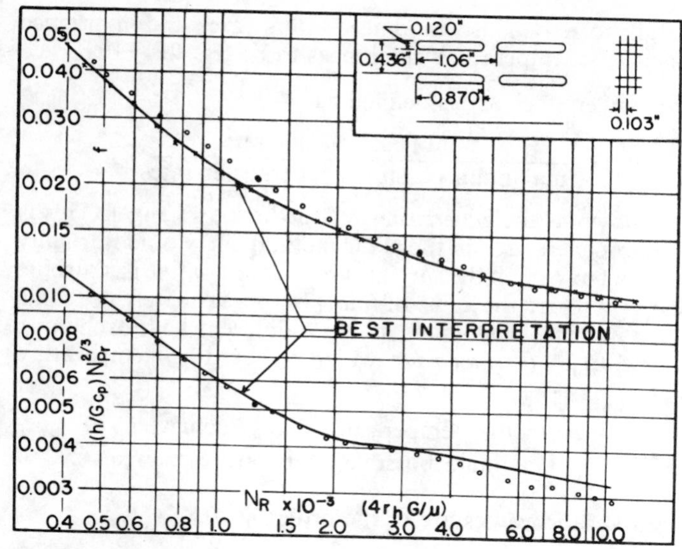

Fin pitch = 9.68 per in.

Flow passage hydraulic diameter, $4r_h = 0.01180$ ft

Fin metal thickness = 0.004 in., copper

Free-flow area/frontal area, $\sigma = 0.697$

Total heat transfer area/total volume, $\alpha = 229\ \text{ft}^2/\text{ft}^3$

Fin area/total area = 0.795

Figure 7.5 Reproduced with permission from Kays, W.M. and London, A.L., *Compact Heat Exchangers*, McGraw-Hill (1964)

7.5 Pressure loss in heat exchangers

Heat exchangers often have complex flow passages leading to excessive pressure loss in the flow. Heat exchanger design requirements may well include a limit to the allowable pressure loss so that an apparently successful heat transfer design may be unacceptable for this reason.

7.6 Efficiency of heat exchangers

The driving potential for heat transfer is the temperature difference. In a heat exchanger this temperature difference varies along the heat exchange surface (Figure 7.2) and it can be seen that the minimum temperature difference will occur at one end of the exchanger. The maximum possible heat transfer rate would occur when this minimum falls to zero. This would need infinite area and in practice it is not usual for temperature differences to be less than 10 K.

Heat exchanger efficiency is defined as

$$E = \frac{\text{actual heat transfer rate}}{\text{maximum possible heat transfer rate}}$$

and this is known as the *effectiveness* of the heat exchanger. This is one of the parameters used in the NTU method and would normally have a value of 0.6–0.8. In the special case of regenerative gas turbine heat exchangers where the value of ($\dot{m}c_p$) is approximately the same for both streams it is customary to use a different heat exchanger efficiency known as *thermal ratio* which is based on temperature change:

$$\text{Thermal ratio} = \frac{\text{actual temperature change achieved}}{\text{maximum possible temperature change}}$$

Again, thermal ratio values are of the order 0.6–0.8.

7.7 Fouling in heat exchangers

The heat transfer rate achieved by a particular heat exchanger depends on the U value. In many examples the thermal resistance of thin metal tube walls is negligible in comparison with the thermal resistance to boundary convection. Although this is true for clean heat exchange surfaces, deposit of sludge or scale from either fluid will produce thermal resistances which are not negligible and design must account for the fall in performance to be expected. To counteract this effect velocities in tubes are often limited to a minimum of about 1 m/s and regular cleaning of heat exchangers is required.

References

1. Bowman, R.A., Mueller, A.C. and Nagle, W.M., 'Mean temperature difference in design', *Transactions of the American Society of Mechanical Engineers*, **62**, 283–94 (1940).
2. Kern, D.Q., *Process Heat Transfer*, McGraw-Hill (1984).
3. Kays, W.M. and London, A.L., *Compact Heat Exchangers*, McGraw-Hill (1964).

WORKED EXAMPLES

Program 7.1 Heat exchanger design considerations

When a heat exchanger is to be designed there will be a number of options to be considered. This simple program attempts to show some of the problems confronting the designer.

Waste heat is to be transferred from a fluid at a constant temperature of 110 °C to water flowing inside a smooth circular tube. The water enters at 50 °C, it is to leave at 80 °C and there is a heat transfer rate of 25 kW. The heat transfer coefficient on the outside of the tube is expected to be 2000 W/m² K. Calculate the tube length required using 1, 2, 3, 4 and 5 cm diameter tubes. Having established these single tube solutions consider a multitube situation using five tubes, 1 cm in diameter, and again find the tube length required for the 25 kW heat transfer rate. The necessary reading for this program is Sections 7.1, 7.2 and 7.5.

To distinguish the advantages and disadvantages of the options the Reynolds number, surface heat transfer coefficient, overall heat transfer coefficient per unit length, heat transfer area and pressure drop will be determined as well as the tube length.

The friction coefficient to determine pressure drop uses the Blasius equation for smooth tubes (equation (4.3)):

$$f = 0.079\ \mathrm{Re}^{-1/4}$$

The water mass flow rate is given by

$$\dot{m} = \frac{\dot{Q}}{\dot{m}c_p}$$

Fluid properties for use in the relation:

$$\mathrm{Nu} = 0.023\ \mathrm{Re}^{0.8}\mathrm{Pr}^{0.4}$$

which may be assumed to be valid if Re > 4000 are determined at the mean bulk temperature.

```
10  PRINT "WASTE HEAT EXCHANGER"
20  REM  THE PROGRAM USES THE LAGRANGE INTERPOLATION FOR
WATER PROPERTIES
30  DIM T(11),VS(11),CP(11),K(11),MU(11),PR(11)
40  FOR N = 0 TO 10
50  READ T(N),VS(N),CP(N),K(N),MU(N),PR(N)
60  NEXT N
70  REM  PUT IN THE PROBLEM DATA
80 Q = 25:T1 = 50:T2 = 80:T3 = 110:H1 = 2000
90  REM  MEAN BULK TEMPERATURE FOR FLUID PROPERTIES
100 T = (T1 + T2) / 2
110 A = T
120  GOSUB 1000
130  REM  MASS FLOW RATE
140 M = Q / (CP * (T2 - T1))
150  REM  CALCULATION LOOP:REYNOLDS NUMBER,HEAT TRANSFER
COEFFICIENT,U VALUE,LENGTH,AREA,PRESSURE DROP
153  PRINT
155  PRINT " D    RE     H      U     L     AR    DP"
156  PRINT
```

```
157  PRINT " M          W/M^2K W/M^2K  M    M^2   N/M^2"
158  PRINT
160  FOR D = 1 TO 5
170 RE = 4 * M * 100 / (3.142 * D * MU)
190 H2 = (K * (RE ^ 0.8) * (PR ^ 0.4) * 0.023 * 100) / D
200 Z = ((1 / H1) + (1 / H2)) * (100 / (3.142 * D))
210 U = 1 / Z
220 TH = ((T3 - T2) - (T3 - T1)) / ( LOG ((T3 - T2) / (T3
 - T1)))
230 L = Q * 1000 / (TH * U)
240 AR = (3.142 * D * L) / 100
250 FC = 0.079 / (RE ^ 0.25)
260 DP = (32 * FC * L * (M ^ 2)) / ((1 / VS) * (3.142 ^ 2
) * ((D / 100) ^ 5))
265 RE =  INT (RE):H2 =  INT (H2):U =  INT (U):L =  INT (
L * 10 + 0.5) / 10:AR =  INT (AR * 100 + 0.5) / 100:DP =
 INT (DP)
270  PRINT  TAB( 2)D; TAB( 5)RE; TAB( 12)H2; TAB( 21)U; T
AB( 26)L; TAB( 32)AR; TAB( 38)DP
280  NEXT D
290  END
1000 B = 0:C = 0:E = 0:F = 0:G = 0
1010  FOR J = 0 TO 10
1020 X = 1
1030  FOR N = 0 TO 10
 |
 ↓

2080  DATA  80,0.00103,4.197,0.67,0.35,2.19
2090  DATA  90,0.00104,4.205,0.676,0.311,1.93
2100  DATA  100,0.00104,4.216,0.681,0.278,1.273
```

```
]RUN
WASTE HEAT EXCHANGER

 D    RE      H       U      L      AR     DP

CM           W/M^2K W/M^2K   M     M^2    N/M^2

 1  59083  14842     55    10.4   .33    69355
 2  29541  4262      85    6.8    .42    1668
 3  19694  2054      95    6      .57    217
 4  14770  1224      95    6.1    .76    55
 5  11816  819       91    6.3    .99    20
```

```
10  PRINT "MULTITUBE EXAMPLE"
15  PRINT
20  REM  OBTAIN FLUID PROPERTIES
30 VS = 0.001022:CP = 4.187:K = 0.657:MU = 0.000429:PR =
2.73
40  PRINT " CHOOSE 5 TUBES OF 1 CM DIAMETER"
50 D = 0.01:N = 5
60 Q = 25:QR = Q / N:T1 = 50:T2 = 80:T3 = 110:H1 = 2000
70 MR = QR / (CP * (T2 - T1))
80 RE = 4 * MR / (3.142 * D * MU)
90 H2 = (K * (RE ^ 0.8) * (PR ^ 0.4) * 0.023) / D
100 Z = ((1 / H1) + (1 / H2)) * (1 / (3.142 * D))
110 U = 1 / Z
120 TH = ((T3 - T2) - (T3 - T1)) /  LOG ((T3 - T2) / (T3
- T1))
130 LR = QR * 1000 / (TH * U):L = 5 * LR
140 FC = 0.079 / (RE ^ 0.25)
150 DP = (32 * FC * LR * (MR ^ 2)) / ((1 / VS) * (3.142 ^
 2) * (D ^ 5))
```

```
160  PRINT " LENGTH OF HEAT EXCHANGER IS ";LR;" M"
170  PRINT " TOTAL LENGTH OF TUBE REQUIRED IS ";L;" M"
180  PRINT " PRESSURE DROP IS ";DP;" N/M^2"

]RUN
MULTITUBE EXAMPLE

 CHOOSE 5 TUBES OF 1 CM DIAMETER
 LENGTH OF HEAT EXCHANGER IS 2.73754944 M
 TOTAL LENGTH OF TUBE REQUIRED IS 13.6877472 M
 PRESSURE DROP IS 1088.88891 N/M^2
```

Program nomenclature

T	water temperature for properties
VS, CP, K, MU, PR	water properties
Q	heat transfer rate
T1	water inlet temperature
T2	water outlet temperature
T3	fluid temperature
H1	surface heat transfer coefficient on outside of tube
M	water mass flow rate
D	tube diameter
RE	Reynolds number
H2	surface heat transfer coefficient inside the tube
U	overall heat transfer coefficient per unit length
TH	log mean temperature difference
L	tube length
AR	heat transfer area
FC	friction coefficient in tube
DP	pressure drop in tube

Program notes

(1) This first program incorporates Program 4.1 for water properties in lines 110 and 1000–2100 for which the temperature is found in line 100.
(2) The remaining structure of the first program is as follows:

Line 140 water mass flow rate.
Lines 155–158 table headings.
Lines 160–280 calculation loop for Reynolds number, surface heat transfer coefficient, overall heat transfer coefficient per unit length, log mean temperature difference, tube length, area and pressure drop.

(3) The results which would all transfer energy at a rate of 25 kW display the following characteristics:

1. At small diameters the pressure drop and length are excessive.
2. There is a minimum length. If we ignore the effect of radius in the relation

$$\frac{1}{U'} = \Sigma \frac{1}{2\pi rh} = \frac{1}{2\pi r} \Sigma \frac{1}{h}$$

the values of h must be approximately equal to obtain matching of the heat transfer coefficients. There is no advantage in the 1 cm result which has a high heat transfer coefficient inside the tube because the dominant smaller outside heat transfer coefficient gives a poor $\Sigma\ 1/h$. When this is combined with small diameter the overall heat transfer coefficient per unit length is poor demanding a long tube and high pressure loss with high pumping costs. As diameter increases, some compensation for the inside heat transfer coefficient being smaller than the outside value is obtained and the overall heat transfer coefficient per unit length does not fall quickly. Design might be for 3, 4 or 5 cm tubes. The minimum cost of the heat exchanger might be at 3 cm diameter since the tube area is least. A balance could be drawn between capital cost (area) and running cost (pressure drop) (see Problem 7.12).

(4) The second program considers a multitube example to achieve the same heat transfer rate. The amendments include the following:

(a) heat transfer rate per tube QR (line 60);
(b) mass flow rate per tube MR (line 130);
(c) length per tube LR (line 130);
(d) total tube length used L (line 130).

(5) With the second program a comparison with the single 1 cm tube and the multitube 1 cm tube can be made as follows:

1. The pressure drop is cut from 69 355 N/m^2 to 1089 N/m^2.
2. The heat exchanger length is cut from 10.4 m to 2.74 m.
3. The total length of tube needed is *increased* from 10.4 m to 13.7 m.

(6) Multitube heat exchangers are commonly used to reduce pressure drop.
(7) Further investigations could be considered by amending the second program and tabulating the results to include more information as in the first program.

Program 7.2 Shell and tube heat exchanger – LMTD method

A single shell pass, two tube pass heat exchanger cools 3 kg/s of oil entering at 200 °C and leaving at 100 °C. Water is used as a coolant entering the tubes at 15 °C and leaving at 65 °C. Write a program to determine the heat exchanger area required and by selecting a tube diameter and length determine the number of tubes required. (The wall resistance and thickness may be assumed to be negligible.) The necessary reading for this program is Section 7.2.

An estimated *U* value (overall heat transfer coefficient) is required for the first part of the program. This enables area to be determined and hence the number of tubes for any chosen size. Using the diameter of the tube the actual *U* value can be determined and compared with the estimate. To achieve this calculation it is assumed that the baffles in the shell give an effective cross-flow velocity of 0.1 m/s. It will also be assumed that provided Re > 4000 in the tubes the relation

$$\mathrm{Nu} = 0.023\ \mathrm{Re}^{0.8}\mathrm{Pr}^{0.4}$$

may be used and that for tube bundles in cross-flow the simple relation

$$\mathrm{Nu} = 0.5\ \mathrm{Re}^{0.5}\mathrm{Pr}^{0.37}$$

may be used. In both cases properties are at the fluid bulk temperature and the Nu are average values.

```
10  PRINT "SHELL AND TUBE HEAT EXCHANGER"
15  PRINT
20  REM  FIRST DETERMINE AREA USING AN ASSUMED U VALUE
30  REM  WATER AND OIL PROPERTIES AT MEAN BULK TEMPERATUR
ES OF 40 C AND 150 C RESPECTIVELY
40 VW = 0.00101:CW = 4.179:KW = 0.632:MW = 0.651:PW = 4.3
1
50 VO = 0.00121:CO = 2.35:KO = 0.134:MO = 10:PO = 152
60  REM  THE LMTD CORRECTION FACTOR COULD BE OBTAINED BY
THE EQUATION TO THE CHART HERE IT IS GIVEN DIRECT
70 F = 0.91
80  REM  F BASED ON COUNTER FLOW VALUE OF LMTD
90 LMTD = ((200 - 65) - (100 - 15)) /  LOG ((200 - 65) /
(100 - 15))
100  REM  OIL FLOW IS 3 KG/S COOLED FROM 200 C TO 100 C
110 Q = 3 * CO * (200 - 100)
120  PRINT "ESTIMATED U VALUE IS 400 W/M^2 K"
125  PRINT
130 U = 400
140 A = Q * 1000 / (U * F * LMTD)
150 X = ( INT (A * 10 + 0.5)) / 10
160  PRINT "HEAT TRANSFER AREA IS ";X;" M^2"
165  PRINT
170  REM  WE NOW SELECT AS AN EXAMPLE TUBES 25 MM IN DIAM
ETER AND 4 M LONG AND FIND THE REQUIRED NUMBER WITH TWO T
UBE PASSES
180 N = A / (2 * 3.142 * 0.025 * 4)
```

```
190 N = 1 +  INT (N)
200  PRINT "NUMBER OF TUBES IS ";N
210  PRINT
1000  REM  SECOND PART OF PROGRAM DETERMINES U VALUE WITH
 THE CHOSEN TUBE SIZE A CROSS FLOW OIL VELOCITY IS ESTIMA
TED AS 0.1 M/S
1010  REM  HEAT TRANSFER COEFFICIENT FOR WATER IN TUBES
1020  REM  WATER FLOW RATE PER TUBE
1025 MFW = Q / (CW * (65 - 15) * N)
1030  REM  REYNOLDS NUMBER
1035 RWE = 1000 * 4 * MFW / (3.142 * 0.025 * MW)
1040  PRINT " WATER REYNOLDS NUMBER IS ";RWE
1045  IF RWE < 4000 THEN  GOTO 1200
1050 NWU = 0.023 * (RWE ^ 0.8) * (PW ^ 0.4)
1055 HW = (NWU * KW) / 0.025
1060  PRINT "WATER SIDE HEAT TRANSFER COEFFICIENT IS ";HW
;" W/M^2 K"
1065  REM  HEAT TRANSFER COEFFICIENT IN SHELL FOR OIL IN
CROSSFLOW
1070 ROE = ((1 / VO) * 0.10 * 0.025 * 1000) / MO
1075  REM  CROSSFLOW NU-RE RELATION
1080 NOU = 0.5 * (ROE ^ 0.5) * (PO ^ 0.37)
1090 HO = (NOU * KO) / 0.025
1095  PRINT "OIL SIDE REYNOLDS NUMBER IS ";ROE
1100  PRINT "OIL SIDE HEAT TRANSFER COEFFICIENT IS ";HO;"
 W/M^2 K"
1110  REM  ASSUME WALL RESISTANCE IS ZERO
1120 Y = (1 / HW) + (1 / HO)
1130 U = 1 / Y
1135 U = ( INT (U * 10 + 0.5)) / 10
1140  PRINT
1150  PRINT "THE U VALUE FOR THE CHOSEN TUBE SIZE "
1160  PRINT "IS ";U;" W/M^2 K COMPARED WITH THE ESTIMATED
 "
1170  PRINT "VALUE OF 400 W/M^2 K.ITERATION IS REQUIRED"
1180  GOTO 1210
1200  PRINT "FLOW NOT TURBULENT IN WATER TUBES"
1210  END
```

```
]RUN
SHELL AND TUBE HEAT EXCHANGER

ESTIMATED U VALUE IS 400 W/M^2 K

HEAT TRANSFER AREA IS 17.9 M^2

NUMBER OF TUBES IS 29

 WATER REYNOLDS NUMBER IS 9100.84163
WATER SIDE HEAT TRANSFER COEFFICIENT IS 1533.06673 W/M^2 K
OIL SIDE REYNOLDS NUMBER IS 206.61157
OIL SIDE HEAT TRANSFER COEFFICIENT IS 247.169137 W/M^2 K

THE U VALUE FOR THE CHOSEN TUBE SIZE
IS 212.9 W/M^2 K COMPARED WITH THE ESTIMATED
VALUE OF 400 W/M^2 K.ITERATION IS REQUIRED
```

Program nomenclature

VW, CW, KW, MW, PW	water properties
VO, CO, KO, MO, PO	oil properties
F	LMTD correction factor

Program notes

LMTD counter flow log mean temperature difference
Q heat transfer rate
U overall heat transfer coefficient
A area of heat exchanger surface
N number of tubes
MFW mass flow rate per tube
RWE water Reynolds number
NWU water Nusselt number
HW water side heat transfer coefficient
ROE oil Reynolds number
NOU oil Nusselt number
HO oil side heat transfer coefficient

(1) The first part of the program is straightforward (lines 10–160) when the equation

$$\dot{Q} = UAF\theta_{\mathrm{LMTD}}$$

is used to find A with an assumed U, an F from the chart (Figure 7.3) and a calculated log mean temperature difference.
(2) In line 170 a tube diameter of 0.025 mm and length of 4 m is chosen and in line 180 the number of tubes to satisfy the area needs is found.
(3) The rest of the program is also straightforward. First the mass flow rate of water in each tube is found (line 1025) and hence the tube Reynolds number (line 1035) to enable the water side heat transfer coefficient to be established (lines 1045–1050). A similar process for the oil side enables the oil side heat transfer coefficient to be found (lines 1070–1100). The U value is then determined for the chosen diameter and length.
(4) If it is clear that the actual U value is much less than the estimate and the program could be rearranged to converge by repeatedly making the estimated U equal to the calculated U.
(5) If the Reynolds number ($4\dot{m}/\pi d\mu$) on the water side falls below 4000 the tube size is too big for turbulent flow and the program ends.
(6) The value of F in line 70 may be obtained from the analytical relation from which the chart was drawn:

$$F = \frac{B\ln[(1 - P)/(1 - RP)]}{(R - 1)\ln\left[\dfrac{2 - P(R + 1 - B)}{2 - P(R + 1 + B)}\right]}$$

where $B = (1 + R^2)^{1/2}$

$$R = \frac{T_1 - T_2}{t_2 - t_1} \quad \text{and} \quad P = \frac{t_2 - t_1}{T_1 - t_1}$$

(see Figure 7.3). This relation could be added as a subroutine entered by changing line 70.

Program 7.3 Cross-flow heat exchanger – NTU method

A cross-flow heat exchanger consists of a series of plates 0.7 m by 0.4 m and 1 mm thick with a gap between of 7 mm. The total number of passages is 41 and of these 21 pass air and 20 pass gas (Figure 7.6). Exhaust gases from a gas turbine enter at 880 °C with a flow rate of 0.62 kg/s and heat 0.78 kg/s of air entering at 20 °C. The relevant NTU chart is shown in Figure 7.4. Determine the temperature of the gas and air at exit from the heat exchanger and the heat transfer rate achieved. Assume that the exhaust gas properties are the same as air properties at the same temperature. The necessary reading for this program is Section 7.3.

These are non-circular flow passages for which the hydraulic diameter is

$$\left[\frac{4 \times \text{cross-sectional area}}{\text{wetted perimeter}}\right]$$

(see Section 4.4). It will be necessary to assume a mean bulk temperature for air and gas and, if required, iterate the solution. In such

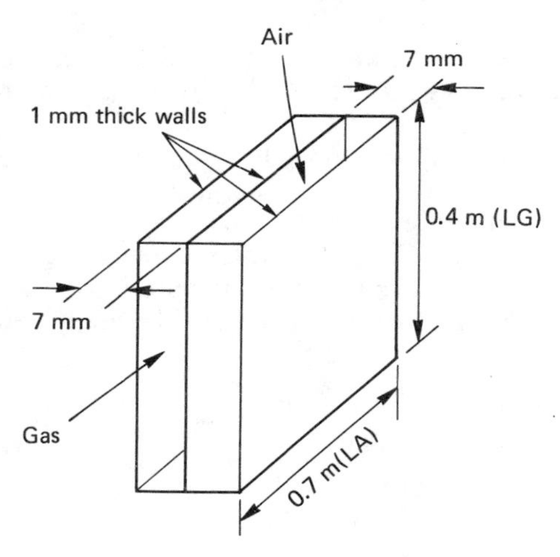

Figure 7.6

a short heat exchanger it is likely that $L/d < 60$ so that an entry length equation will be required to determine surface heat transfer coefficient. For this situation:

$$\mathrm{Nu} = 0.036\,\mathrm{Re}^{0.8}\mathrm{Pr}^{0.33}\left(\frac{d}{L}\right)^{0.055}$$

which is assumed valid if Re > 3000.

```
10  PRINT "CROSS FLOW HEAT EXCHANGER"
20  REM  FLUID PROPERTIES AT ESTIMATED MEAN BULK TEMPERAT
URES OF 200 C FOR AIR AND 700 C FOR GAS(AIR PROPERTIES)
30 VA = 1.341:CA = 1.03:KA = 0.039:MA = 0.000026:PA = 0.6
9
40 VG = 2.756:CG = 1.14:KG = 0.066:MG = 0.000042:PG = 0.7
3
50  REM  EVALUATE THE HYDRAULIC DIAMETERS
60 LG = 0.4:LA = 0.7:W = 0.007
70 DA = 4 * W * LA / (2 * (W + LA))
80 DG = 4 * W * LG / (2 * (LG + W))
90  REM  MASS FLOW RATES FOR REYNOLDS NUMBERS
100 M1A = 0.78:M2G = 0.62
110 R1E = M1A * DA / (21 * W * LA * MA)
120 R2E = M2G * DG / (20 * W * LG * MG)
122  PRINT "REYNOLDS NUMBER-AIR ";R1E;" L/D-AIR= ";LG / D
A
124  PRINT "REYNOLDS NUMBER-GAS= ";R2E;" L/D-GAS= ";LA /
DG
126  IF R1E < 3000 THEN  GOTO 430: IF R2E < 3000 THEN  GO
TO 430
128  IF LA / DG > 60 THEN  GOTO 430: IF LG / DA > 60 THEN
 GOTO 430
130  REM   EVALUATE SURFACE HEAT TRANSFER COEFFICIENTS FO
R ENTRY LENGTH TURBULENT FLOW AS L/D RATIO IN EACH CASE I
S LESS THAN 60
140 HA = (0.036 * (R1A ^ 0.8) * (PA ^ 0.33) * ((DA / LG)
^ 0.055) * KA) / DA
150 HG = (0.036 * (R2G ^ 0.8) * (PG ^ 0.4) * ((DG / LA) ^
 0.055) * KG) / DG
160  REM  EVALUATE THE NUMBER OF TRANSFER UNITS
170 A = LG * LA * 2 * 20
180 UA = 1 / ((1 / (HA * A)) + (1 / (HG * A)))
190 C1A = M1A * CA
200 C2G = M2G * CG
210  IF C1A > C2G THEN  GOTO 230
220 NTU = UA / (1000 * C1A): GOTO 240
230 NTU = UA / (1000 * C2G)
240  REM  EVALUATE THE CAPACITY RATIO
250 X = C1A / C2G
260  IF X > 1 THEN X = 1 / X
270  PRINT "ENTER THE EFFECTIVENESS CHART WITH NTU= ";NTU

280  PRINT " AND CAPACITY RATIO ";X
290  PRINT "ENTER EFFECTIVENESS": INPUT E: PRINT "EFFECTI
VENESS IS ";E
300  REM  FIND AIR AND GAS EXIT TEMPERATURES AND HEAT TRA
NSFER RATE
310  REM  INLET TEMPERATURES
320 TG = 880:TA = 20
330  IF C1A > C2G THEN  GOTO 360
340 TNA = TA + E * (TG - TA)
350 TMG = TG - (C1A * (TNA - TA)) / C2G: GOTO 380
```

```
360 TMG = TG - E * (TG - TA)
370 TNA = TA + (C2G * (TG - TMG)) / C1A
380 Q = C2G * (TG - TMG)
390  PRINT
400  PRINT "AIR OUTLET TEMPERATURE IS ";TNA;" C": PRINT
410  PRINT "GAS OUTLET TEMPERATURE IS ";TMG;" C": PRINT
420  PRINT "HEAT TRANSFER RATE IS ";Q;" KW"
425  END
430  PRINT "HEAT TRANSFER COEFFICIENT EQUATION INVALID":
END

]RUN
CROSS FLOW HEAT EXCHANGER
REYNOLDS NUMBER-AIR 4041.22045 L/D-AIR= 28.8571429
REYNOLDS NUMBER-GAS= 3627.00363 L/D-GAS= 50.875
ENTER THE EFFECTIVENESS CHART WITH NTU= .545186188
 AND CAPACITY RATIO .879761015
ENTER EFFECTIVENESS
?0.375
EFFECTIVENESS IS .375

AIR OUTLET TEMPERATURE IS 303.722927 C

GAS OUTLET TEMPERATURE IS 557.5 C

HEAT TRANSFER RATE IS 227.943 KW
```

Program nomenclature

VA, CA, KA, MA, PA	air properties
VG, CG, KG, MG, PG	gas properties
LG	gas passage size
LA	air passage size
W	passage width
DA	air hydraulic diameter
DG	gas hydraulic diameter
M1A	air mass flow rate
M2G	gas mass flow rate
R1E	air Reynolds number
R2E	water Reynolds number
HA	air surface heat transfer coefficient
HG	gas surface heat transfer coefficient
A	heat transfer area
NTU	number of transfer units
X	capacity ratio
E	effectiveness
TG	gas inlet temperature
TA	air inlet temperature
TNG	gas outlet temperature
TMA	air outlet temperature
Q	heat transfer rate.

Program notes

(1) The property value Program (4.1) with air data could be incorporated.
(2) The equation to the effectiveness chart could be used as a subroutine instead of line 290:

$$E \simeq 1 - \exp\left[\frac{NTU^{0.22}}{C}\{\exp(-C \cdot NTU^{0.78}) - 1\}\right]$$

where C is the capacity ratio and NTU is the number of transfer units. Clearly the chart is difficult to read at low NTU values.
(3) The program structure is straightforward, as follows:

Lines 10–40 gas and air properties.
Lines 50–80 hydraulic diameters.
Lines 90–124 Reynolds numbers.
Lines 126–128 checking equation validity.
Lines 140–150 surface heat transfer coefficients.
Lines 170–180 area and overall heat transfer coefficient.
Lines 190–280 capacity ratio and NTU values to find effectiveness.
Line 290 using the chart (see note 2) and entering effectiveness.
Lines 300–370 finding exit temperatures (see note 4).
Line 38 determination of Q.
Line 430 escape line if Re < 3000 or $L/d > 60$.

(4) The definition of E (equation (7.5)):

$$E = \frac{(\dot{m}c_p\Delta T)_{hot}}{(\dot{m}c_p)_{min}(T_{hot_{in}} - T_{cold_{in}})} = \frac{(\dot{m}c_p\Delta T)_{cold}}{(\dot{m}c_p)_{min}(T_{hot_{in}} - T_{cold_{in}})}$$

is simplified in lines 300–370. If, for example, $(\dot{m}c_p)_{min}$ is $(\dot{m}c_p)_{hot}$ then

$$E = \frac{\Delta T_{hot}}{(T_{hot_{in}} - T_{cold_{in}})} \quad \text{and} \quad T_{hot_{out}} = T_{hot_{in}} - E(T_{hot_{in}} - T_{cold_{in}})$$

and $T_{cold_{out}}$ is found by

$$(\dot{m}c_p\Delta T)_{cold} = (\dot{m}c_p\Delta T)_{hot}$$

Alternatively, if $(\dot{m}c_p)_{min}$ is $(\dot{m}c_p)_{cold}$ then

$$T_{cold_{out}} = T_{cold_{in}} + E(T_{hot_{in}} - T_{cold_{in}})$$

and $T_{hot_{out}}$ is found by

$$(\dot{m}c_p\Delta T)_{hot} = (\dot{m}c_p\Delta T)_{cold}$$

This point is important in the application of the NTU method.
(5) The actual mean air temperature is 160 °C and the mean gas temperature is 719 °C so that an iteration might be used to check the result.
(6) Clearly this program has been written for this problem and is not general. By altering a few lines it should be possible to test the program for different geometric arrangements to see if a more effective heat exchanger could be designed. The chart (Figure 7.4) shows that the effectiveness is increasing quickly at the point of operation and if the number of transfer units could be increased to about 2 the effectiveness could be doubled (Problem 7.6).

Program 7.4 Car radiator design

When heat exchangers have complex geometry and an analytical relation of the form Nu $= \phi$(Re, Pr) is not available the manufacturers supply experimental data in graphical form. This data also includes other important information. Figure 7.5 shows a chart. It is proposed to use this matrix for a car radiator. The material has a thermal conductivity of 390 W/m K and the radiator is to be 0.3 m wide, 0.4 m high and 2.12 inches (0.053 85 m) thick (see Figures 7.5 and 7.7). The air mass flow rate through the radiator will be 0.7 kg/s. The fin thickness from Figure 7.5 is 0.004 inches which may be considered as 0.1 mm. The water flow rate is 0.3 kg/s. The air enters the radiator at 15 °C and the water enters at 100 °C.

Write a program to determine the air and water exit temperatures and the heat transfer rate achieved in the radiator.

The NTU method is chosen as exit temperatures are unknown and the problem is one of cross-flow with unmixed fluids so that Figure 7.4 is required to obtain the effectiveness. The fins on the air side will

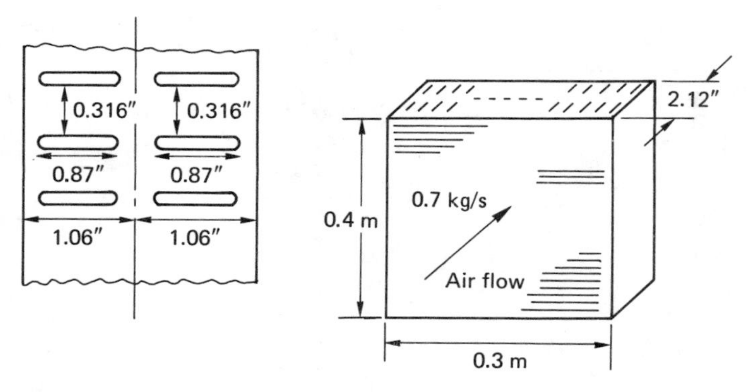

Figure 7.7

require the area weighted fin efficiency method of Section 6.4. It will be seen that the data on Figure 7.5 for the matrix includes the ratio β = fin area/total area and the total heat transfer area is obtained from the parameter α given on the chart (α = total heat transfer area/unit volume). The hydraulic diameter of the surface is also given on the chart to enable the air side Reynolds number to be found. It is quite clear that the manufacturer of the matrix supplies the essential data for the complex surface. The symbol G represents the mass flow rate per unit (of free flow) area.

The equation for surface heat transfer coefficient on the water side for turbulent flow if $L/d > 60$, $\mathrm{Re} > 3000$ is

$$\mathrm{Nu} = 0.023\ \mathrm{Re}^{0.8}\mathrm{Pr}^{0.4}$$

If these conditions are not valid the program must stop. The necessary reading for this program is Sections 7.3, 7.4 and 6.4.

```
10  PRINT "COMPACT HEAT EXCHANGER": PRINT
20  REM  ASSUME MEAN BULK TEMPERATURE OF AIR IS 25 C AND
OF WATER IS 80 C FOR PROPERTIES
30 VW = 0.00103:CW = 4.197:KW = 0.67:MW = 0.00035:PW = 2.
19
40 VA = 0.844:CA = 1.013:KA = 0.026:MA = 0.000017:PA = 0.
705
50  REM  CALCULATE HEAT TRANSFER COEFFICIENTS FOR WATER T
UBES THEN AIR FINNED MATRIX
60  REM  WATER SIDE
70 M1W = 0.3:T = 0.05385:W = 0.3:L = 0.4
80  REM  TUBE HYDRAULIC DIAMETER IS 4*AREA/PERIMETER
90 DT = 2.54 * (4 * 0.87 * 0.12) / (2 * (0.87 + 0.12))
95  REM  NUMBER OF TUBES
100 N = (2 * 0.3 * 100) / (0.436 * 2.54)
110 R1E = (4 * M1W * 100) / (N * 3.142 * DT * MW)
120 X = (L * 100) / DT
124  IF R1E < 3000 THEN  GOTO 128
126  IF X < 60 THEN  GOTO 128
127  GOTO 130
128  PRINT "HEAT TRANSFER EQUATION IN LINE 130 INVALID":
END
130 HW = (0.023 * KW * (R1E ^ 0.8) * (PW ^ 0.4) * 100) /
DT
140  REM  AIR SIDE
150  PRINT " FROM MATRIX DIAGRAM OBTAIN HYDRAULIC DIAMETE
R": PRINT "AND FLOW AREA/FRONTAL AREA RATIO"
160  INPUT D2A: INPUT AR
165  PRINT "HYDRAULIC DIAMETER IS ";D2A;"FT AND AREA RATI
O IS ";AR:M2A = 0.7
166  PRINT
170  REM  CHART BASE FOR RE IS G*D/MU WHERE G IS MASS FLO
W PER UNIT AREA
180 DA = D2A * 12 * 2.54
190 R2E = (M2A * DA) / (AR * L * W * MA * 100)
200  PRINT "ENTER MATRIX CHART AT NR*10^-3= ";R2E / 1000:
 PRINT " AND OBTAIN (H/G*CP)NPR^2/3"
210  INPUT ZZ: PRINT "FROM CHART F(RE)= ";ZZ
215  PRINT
```

```
220 HA = (ZZ * 1000 * M2A * CA) / ((PA ^ 0.67) * AR * L *
 W)
230  REM  FIND U VALUE USING FIN EFFICIENCY,FIN LENGTH IS
 HALF TUBE SPACING
240 LF = (0.436 - 0.12) * 2.54 / 200:TT = 0.0001:KF = 390

250 MM = ((2 * HA) / (KF * TT)) ^ 0.5
260 Y = MM * LF
270 EF = (( EXP (Y) -  EXP ( - Y)) / ( EXP (Y) +  EXP ( -
 Y))) / Y
280  PRINT "FIND AIR SIDE AREA-VOLUME RATIO AND ": PRINT
"FIN AREA/TOTAL AREA RATIO FROM CHART"
290  INPUT A: INPUT B: PRINT "A/V RATIO IS ";A;" FT^2/FT^
3,FIN AREA RATIO IS ";B
295  PRINT
300 EX = B * EF + (1 - B)
310  REM  OVERALL HEAT TRANSFER COEFFICIENT BASED ON AIR
SIDE AREA
320  REM  CALCULATE WATER SIDE AREA/VOLUME RATIO
330 AWV = (12 * 2 * (0.12 + 0.87)) / (0.436 * 1.06 * 0.30
48)
340 AAV = A / 0.3048
350 ZX = AWV / AAV
360 U = 1 / ((1 / (EX * HA)) + (1 / (ZX * HW)))
370 UA = U * AAV * W * L * T
380 C1W = M1W * CW * 1000
390 C2A = M2A * CA * 1000
400  IF (C1W > C2A) THEN  GOTO 420
410 NTU = UA / C1W: GOTO 430
420 NTU = UA / C2A
430 S = C1W / C2A: IF S > 1 THEN S = 1 / S
440  PRINT " ENTER THE EFFECTIVENESS DIAGRAM WITH ": PRIN
T "NTU= ";NTU;" AND CAPACITY RATIO ";S
450  INPUT E: PRINT "EFFECTIVENESS= ";E: PRINT
460  REM  INLET TEMPERATURE DATA
470 TA = 15:TW = 100
480  IF (C1W > C2A) THEN  GOTO 510
490 TMW = TW - E * (TW - TA)
500 TNA = TA + (C1W * (TW - TMW)) / C2A:GOTO 530
510 TNA = TA + E * (TW - TA)
520 TMW = TW - (C2A * (TNA - TA)) / C1W
530 Q = C1W * (TW - TMW) / 1000
540  PRINT "AIR OUTLET TEMPERATURE IS ";TNA;" C"
550  PRINT "WATER OUTLET TEMPERATURE IS ";TMW;" C": PRINT

560  PRINT "HEAT TRANSFER RATE IS ";Q;" KW"
570  END


]RUN
COMPACT HEAT EXCHANGER

 FROM MATRIX DIAGRAM OBTAIN HYDRAULIC DIAMETER
AND FLOW AREA/FRONTAL AREA RATIO
?0.0118
?0.697
HYDRAULIC DIAMETER IS .0118FT AND AREA RATIO IS .697

ENTER MATRIX CHART AT NR*10^-3= 1.77064731
 AND OBTAIN (H/G*CP)NPR^2/3
?0.0044
FROM CHART F(RE)= 4.4E-03

FIND AIR SIDE AREA-VOLUME RATIO AND
FIN AREA/TOTAL AREA RATIO FROM CHART
?229
```

```
?0.795
A/V RATIO IS 229 FT^2/FT^3,FIN AREA RATIO IS .795

 ENTER THE EFFECTIVENESS DIAGRAM WITH
NTU= .297817951 AND CAPACITY RATIO .563180049
?0.27
EFFECTIVENESS= .27

AIR OUTLET TEMPERATURE IS 37.95 C
WATER OUTLET TEMPERATURE IS 87.0750179 C

HEAT TRANSFER RATE IS 16.273845 KW
```

Program nomenclature

VW, CW, KW, MW, PW	water properties
VA, CA, KA, MA, PA	air properties
M1W	mass flow rate of water
T	radiator thickness
W	radiator width
L	radiator height
DT	water tube hydraulic diameter
N	number of tubes
R1E	waterside Reynolds number
X	length/diameter ratio on water side
HW	water side heat transfer coefficient
DA, D2A	air side hydraulic diameter
AR	free flow area/frontal area ratio
M2A	air mass flow rate
R2E	air side Reynolds number
ZZ	Prandtl number parameter from chart $h/Gc_p \cdot Pr^{2/3}$
HA	air side heat transfer coefficient
PA	air Prandtl number
LF	fin length
TT	fin thickness
KF	fin thermal conductivity
MM	fin parameter $(ph/ka)^{1/2} = (2h/kt)^{1/2}$
Y	fin parameter ml
EF	fin efficiency (tanh ml)/ml
A	heat transfer area/unit volume
B	fin area/total area
EX	area weighted fin efficiency
AWV	water side area volume ratio
AAV	air side area volume ratio
ZX	A_2/A in equation (6.6)
U	overall heat transfer coefficient based on air side area
C1W	$(\dot{m}c_p)$ water value
C2A	$(\dot{m}c_p)$ air value
NTU	number of transfer units
S	capacity ratio
E	effectiveness

TW inlet water temperature
TMW exit water temperature
TNA exit air temperature
Q heat transfer rate

Program notes

(1) The program structure is as follows:

Lines 30 and 40 water and air properties.
Lines 50–130 calculation of water side heat transfer coefficient.
Lines 140–200 calculation of air side Reynolds number.
Lines 210–220 calculation of air side heat transfer coefficient.
Lines 240–300 calculation of area weighted fin efficiency.
Lines 310–370 calculation of UA based on air side area.
Lines 380–410 calculation of NTU and capacity ratio.
Line 440 determination of effectiveness from chart.
Lines 460–520 calculation of outlet temperatures.
Line 530 calculation of heat transfer rate.

(2) As in other problems the program could determine properties from Program 4.1 and determine effectiveness from the equation given in Program 7.3, note 2.
(3) Similarly in lines 200, 210 the determination of $(h/Gc_p)\cdot Pr^{2/3}$ could be achieved either by the interpolation routine (Program 1.1) or by polynomial curve fit.
(4) The radiator will not achieve enough heat transfer in some situations and may need some alterations. Consider the effects of the relevant parameters (size, flow rates, orientation, etc.) and decide on the best way of achieving higher heat transfer rates. Try to achieve the heat transfer required by an engine developing 30 kW of power.
(5) There are many configurations of compact heat exchanger surfaces given in Reference 3 of Chapter 7 and it might be possible to achieve better results with a different geometry.
(6) As in Program 7.3 the effectiveness of 0.27 is low and it should be possible to improve the NTU and double this value by changing the design.

PROBLEMS

(7.1) Enhance the second part of Program 7.1 to include a wider range of single tube and multitube options. Investigate further the relationships between area, length and pressure loss. What happens if the surface heat transfer coefficient on either side increases or decreases due to cleaning or fouling? Is it possible to determine the range of variation in coefficient to keep the heat transfer within 5%

of the desired value? It may be useful to plot graphs of, for example, area vs. diameter and pressure loss vs. diameter.

(7.2) Continue Program 7.2 until convergence of estimated and actual U values is achieved. Then change the chosen tube size to get a better balance of heat transfer coefficients on the outside and inside of the tubes. What happens if the tube length is changed? Include pressure drop calculations in the program. If Problems 7.1 and 7.2 have been fully exploited the tremendous range of options available in heat exchanger design should begin to become apparent. Reading Reference 2 to Problem 7.12 would be beneficial at this stage.

(7.3)

(a) Solve the problem of Program 7.2 by the NTU method when the two exit temperatures are unknown. The mass flow rate of water is 3.374 kg/s. An NTU chart for the shell and tube heat exchanger will be required [1]. (All references in these problems appear together after Problem 7.12.)

(b) A stream of fluid is cooled by water flowing in a single smooth tube with the fluid in counterflow in an annular space outside the tube. The heat transfer coefficient on the fluid side is 3 W/m^2 K. The flow rate of fluid is 0.5 kg/s with an entry temperature of 130 °C and specific heat capacity 2.24 kJ/kg K. The flow rate of water in the tube is 0.3 kg/s with an entry temperature of 10° C.

Write a program to determine the fluid and water exit temperatures and the heat transfer rate for tube diameters of 10–100 mm with a tube length of 10 m, 20 m and 30 m. What are the pressure losses in each case?

For counterflow in a single tube:

$$E = \frac{1 - \exp[-\text{NTU}(1 - C)]}{1 - C\exp[-\text{NTU}(1 - C)]}$$

(7.4) Program 7.3 can be rearranged to be solved by the LMTD method. The number of passages will not be known but the heat transfer requirement is 229 kW. An LMTD correction factor chart will be required for an unmixed flow cross-flow heat exchanger [1]. Iteration of both temperature and U value will be required. Start by guessing the gas outlet temperature at 500 °C and the U value as 20 W/m^2 K.

(7.5) Shell diameters for shell and tube heat exchangers can be *estimated* from equations based on tube pitch arrangements (Figure 4.6). For square pitched tubes with more than 25 tubes ($N > 25$):

$$\frac{D_{\text{shell}}}{D_{\text{tube}}} = 1.37N^{0.475}$$

This relation should be incorporated into Problem 7.2.

(7.6) Program note 6 for Program 7.3 highlighted the low effectiveness of the design. By changing the geometry of the design it is possible to improve the effectiveness to perhaps 60%. This will give higher heat transfer rates or a smaller area heat exchanger. Change the parameters *LG*, *LA* and *W* to achieve this but other changes will be required if the value of '*L/D*' is less than 60.

(7.7) Program note 6 for Program 7.4 also noted low effectiveness. Find a way of improving this and hence using a smaller radiator. Try to write the program so that it can be used for various radiator designs with different sized engines which will need to dissipate energy at different rates.

(7.8) It can be shown that for a single tube of diameter d and length L with fluid flowing in the tube at velocity V, the pressure drop Δp is given by

$$\Delta p = \frac{4f\rho V^2}{2}\left(\frac{L}{d}\right) \qquad (7.6)$$

where f is the friction coefficient and ρ the fluid density. For a single tube with surface heat transfer coefficient h, an energy balance between convection heat transfer and the energy change of the fluid gives

$$h(\pi dL)\theta_{\mathrm{LMTD}} = \dot{m}c_p\Delta T = \left(\frac{\pi}{4}d^2\rho V\right)c_p\Delta T$$

Hence

$$\frac{\Delta T}{\theta_{\mathrm{LMTD}}} = 4\left(\frac{h}{\rho V c_p}\right)\left(\frac{L}{d}\right) = 4\mathrm{St}\left(\frac{L}{d}\right) \qquad (7.7)$$

From equations (7.6) and (7.7)

$$\frac{\Delta T}{\theta_{\mathrm{LMTD}}} = \frac{\mathrm{St}}{f/2}\cdot\frac{\Delta p}{\rho V^2} \qquad (7.8)$$

In equation (7.8), a modified Reynolds analogy may be substituted for the term $\mathrm{St}/(f/2)$ and then if Δp is specified and $\Delta T/\theta_{\mathrm{LMTD}}$ is obtained from heat transfer needs, the velocity V can be found for any diameter d. This enables L to be found from equation (7.7) and hence the number of tubes to achieve the required heat transfer rate. Two problems follow; write a program to solve both:

(a) A counterflow heat exchanger for a heat recovery system uses waste water to preheat air. The air flow rate is 0.08 kg/s at atmospheric pressure along thin metal tubes of diameter d. The pressure drop is limited to p kN/m^2. Calculate the number of tubes in parallel and their lengths for various values of d and p limited to be between 5 and 15 kN/m^2. Equation (4.3) may be used for f and Colburn's modification of Reynolds analogy is valid:

Air inlet temperature	0 °C
Air exit temperature	21 °C
Water inlet temperature	45 °C
Water exit temperature	10 °C

(b) A range of water heaters consists of a smooth tube with electric heating wire wrapped around the outside so that the tube walls are at 90 °C. The tube is of diameter d and the pressure drop should not exceed p kN/m^2. The water is to be heated from 15 °C to 60 °C. It may be assumed that f can be obtained from equation (4.3) and that the Colburn modification of the Reynolds analogy applies. Determine the water flow rates and heater powers for a range of diameters d and pressure losses between 1 and 5 kN/m^2.

Open-ended problems

(7.9) Problem 4.12 studied heat transfer coefficients in tube banks. Reconsider Program 7.2 in the light of the information obtained in Problem 4.12 and replace the simple tube bundle relation with one based on the later work. Try differing arrangements to see what influence this has on heat exchanger size.

(7.10) Figure 7.8 shows the arrangement of an air blast freezer which is required to freeze M tonne/hr of fish. The specific enthalpy change associated with freezing is 450 kJ/kg. The air enters the heat exchanger of the freezer at −30 °C and the secondary refrigerant liquid (non-evaporating) inside the tubes of the heat exchanger

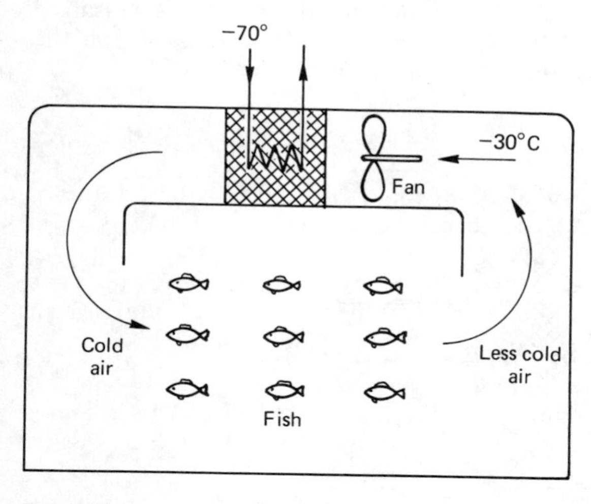

Figure 7.8

enters at $-70\,°C$. The freezer matrix is that used in Program 7.4 but the number of rows of tubes may be up to six. The air velocity at entry to the matrix will depend on the cross section of the flow area. Write a program to enable a suitable sized heat exchanger to be chosen for values of M between 1 and 5 tonne/hr. Assume that the secondary refrigerant has values of ρ, c_p, k and μ identical to those of water at 20 °C.

(7.11) Figure 7.9 is a sketch of the principle of a rotary regenerator in which energy is transferred from a flow of hot gas to a flow of cold gas by a 'thermal wheel' which rotates slowly so that the wheel is heated as it passes through the hot stream and cooled as it passes through the cold stream. It should be possible to model this process by considering the 'mass flow rate' of rotor through each stream in turn.

An NTU type analysis is possible using the charts in Kays and London [1] to validate the results of a problem written using the modelling method suggested above but it is important to read the text relevant to the charts.

Make a study of rotary regenerators for heat transfer and write a program to give approximate solutions for simple problems. Air enters the rotary regenerator fitted between the compressor and combustion chamber of an automobile gas turbine at 350 °C. The air mass flow rate is 1 kg/s. The exhaust gases enter the regenerator at 650 °C at the same mass flow rate. Assume that the properties of exhaust gas are identical to those of air. Determine the heat transfer rate and the exit temperature of the air from the regenerator.

(7.12) Design a horizontal shell and tube heat exchanger using a single shell pass and two tube passes to condense steam on the outside of the tubes containing the cooling water flow [2].

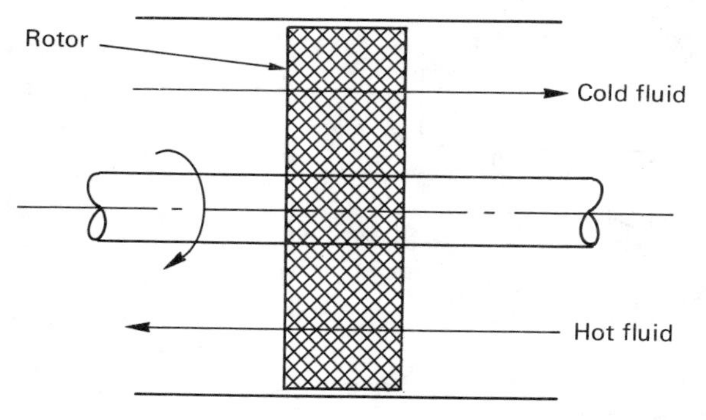

Figure 7.9

Data

Mass flow rate of steam	0.5 kg/s
State of steam	Dry saturated at 40 °C
Coolant entry temperature	10 °C
Minimum heat exchanger length	2 m
Maximum heat exchanger length	4 m

The program should give a range of results to enable a choice to be made and utilise as much previous work as is thought to be relevant. Clearly, one method of determining choice is operating cost, however, this should not include the water charges which may well swamp all other considerations.

Operating cost C may be expressed by an equation of the form

$$C = \text{pumping cost} + \text{cleaning and servicing cost}$$

$$C = Y\frac{\dot{m}\Delta p}{e}\cdot T + X\cdot\frac{A}{D^a}$$

where A is the tube surface area
D is the tube diameter
X is the cleaning and servicing cost factor per unit time
a is an index less than unity
Y is the cost of energy
$\dot{m}$ is the mass flow rate of coolant
Δp is the pressure loss in the tubes
ρ is the water density
and T is the utilisation factor (fraction of the time unit that the heat exchanger operates, $0 \leqslant T \leqslant 1$)

Examine the form of the equation for logic and see if the application of this relation can give a clear choice within the options found by the program. If not, engineering experience will be the dominant factor regarding choice.

Investigate the mechanical aspects of the design of the heat exchanger including structure, stress and vibration problems.

References

1. Kays, W.M. and London, A.L., *Compact Heat Exchangers*, McGraw-Hill (1964).
2. Taborek J., 'Evolution of heat exchanger design techniques', *Heat Transfer Engineering*, **1**(1), 15–29 (1979).

Index